BATTLE LEADERSHIP

Captain Adolf Von Schell

D1027919

ECHO POINT BOOKS & MEDIA, LLC

Copyright © 1933

Published by Echo Point Books & Media,
www.EchoPointBooks.com

ISBN: 978-1-62654-965-4

Cover photograph courtesy of U. S. Marine Corps

Cover design by Adrienne Nunez,
Echo Point Books & Media

Printed in the U.S.A.

Preface

Captain Von Schell's collection of lessons learned as a small unit infantry commander during World War I should be a part of every Marine's professional library. It is one of the finest works of its type I have read and compares favorably with Rommels *Infantry Attacks*.

His observations on combat leadership and tactics are timeless and are as pertinent today as they were in 1917. It should be required reading for all combat leaders, particularly those serving on the platoon, company and battalion level.

D.M. Twomey
Major General, U.S. Marine Corps

Acknowledgement

The author desires to thank Lieutenant Colonel George C. Marshall and the following officers of the Military History Section of The Infantry School, Fort Benning, Georgia, for their encouragement, advice and assistance in the preparation of this book: Major Edwin F. Harding, Major Truman Smith, Major Richard G. Tindall, Captain Francis H. Wilson, Captain John A. Andrews, and First Lieutenant Charles T. Lanham.

Most of this material was originally prepared for informal, oral presentation. It is through the interest of these American friends and their assistance in arranging and editing that these papers are published in their present form.

Foreword

The author of these lectures and articles, Captain Adolf von Schell of the German Army, is a young officer with an extensive and varied combat service.

He entered the Imperial German Army a few years prior to the outbreak of the World War. The mobilization in 1914 found him on the Belgian frontier in command of an infantry platoon of the Westphalian VII Corps.

He participated in the reduction of Liege and in the advance into France which culminated in the battle of the Marne. In October, 1914, he took part in several of the battles which occurred during the historic "race to the sea." In one of these he received the first of several wounds the World War was to bring him, and he was invalided home.

In February, 1915, he returned to duty in East Prussia in command of a newly raised company. There, under Hindenburg, his company marched and fought in meter deep snow in the series of engagements known historically as the Winter Battles of the Mazurian Lakes in which over one hundred thousand prisoners were taken. From July to October of the same year he took part in the gigantic Austro-German assault on the Polish salient, which was climaxed by the fall of Warsaw. Following this, the German drive to the east carried the young officer into Russia as far as the Beresina River. This summer, filled with endless marches and repeated engagements gave him a wide experience in the open warfare tactics of small units.

1916 found Captain von Schell still in the east, operating in close conjunction with the Austrians in an effort to stop the Brussilov steam-roller in Galicia and Wolhynia.

In 1917 his division was transferred to the Rumanian Front where von Schell took part in a number of engagements in the higher altitude of the Carpathians.

In 1918 the war carried the author into its bypaths—the Ukraine, the Crimea, the Caucasus. The Armistice found him operating against the Bolsheviki.

Since the war Captain von Schell has served in many capacities in the new German Army, both in the line and the staff. In 1930 he was designated by his government to attend The Infantry School at Fort Benning and he graduated from that institution with the Advanced Class of 1930-31. Most of the lectures contained in this volume were delivered either to the students of the Advanced and Company Officers' Classes or to the officers of the 29th Infantry.

In presenting these lectures Captain von Schell is placing his experience in open warfare at the disposal of the American Army. Unfortunately, during the last war nearly all of our action fell within the period of stabilization. There is probably no officer in our army who has participated in so many widely varying types of military operations as Captain von Schell. His knowledge of war was gained firsthand in meeting engagements, advance-guard battles, rear-guard actions, night attacks, guerilla warfare, mountain fighting—in which he took part. He has run the entire gamut of tactical experience, from pursuit to withdrawal, from an assault on permanent fortifications to a major offensive in the rigors of a Russian winter. There is no page of his story, as set forth in these lectures, that is not worthy of study by American officers.

CAMPBELL KING,
Major General, U. S. Army,
Commandant, The Infantry School.

Contents

ILLUSTRATIONS

CHAPTER I

Battlefield Psychology

Psychology, as I understand it, means knowledge of the soul. Yet, how shall we speak about the souls of others, when we do not even know our own? Is there a single one of us who can say with certainty how he will react to a given event? Nevertheless, as leaders we must have some knowledge of the souls of our soldiers; because the soldier, the living man, is the instrument with which we have to work in war.

The great commanders of all times had a real knowledge of the souls of their soldiers. Let us use a more simple phrase and call this knowledge of the soul, "knowledge of men." Knowledge of men in all wars has proved an important factor to the leader. It is probable that this will be still more true in future wars.

Prior to the World War, all armies fought in comparatively close order. The psychological reaction of the individual soldier was not so decisive since the fighting was done, not by the individual, but by the mass, and the mass was held together by drill and discipline. Moreover, the psychological impressions of battle were simpler. Rifle and cannon ruled the battlefield, and the enemy could be seen. In modern war, the impressions are much more powerful. Usually we fight against an enemy we cannot see. The machine rules the battlefield. We no longer fight in great masses, but in small groups, often as individuals. Therefore, the psychological reaction of the individual has become increasingly important. As commanders we must know the probable reaction of the individual and the means by which we can influence this reaction.

The knowledge of men is especially difficult for two reasons: first, because it cannot be learned from books; second because the characteristics of the individual in peace are completely changed in war. Man reacts differently in war than he does in peace, therefore he must be handled differently. For this reason we cannot learn, in peace-times, the psychology of war. It is my belief that no one can give a prescription for a correct application of the principle of psychology in war. The only thing of which we are certain is this: the psychology of the soldier is always important. No commander lacking in this inner knowledge of his men can accomplish great things.

As long as armies were small and the battlefield narrow, a leader could exert a psychological influence on his army by personal example. In modern wars, however, the high commanders are necessarily far in rear and the majority of the soldiers never see them. Consequently, the task of influencing and understanding the soldier psychologically has, in large measure, passed to subordinate commanders. For this reason we shall deal only with the psychology of individuals and small units.

In peace we should do everything possible to prepare the minds of our soldiers for the strain of battle. We must repeatedly warn them that war brings with it surprise and tremendously deep impressions. We must prepare them for the fact that each minute of battle brings with it a new assault on nerves. As soldiers of the future, we ourselves should strive to realize that we will be faced in war by many new and difficult impressions; dangers that are thus fore-seen are already half overcome.

Let us take several actual examples from war and see what we can learn from them. In studying these examples it should be borne in mind that they do not constitute a universal formula; they deal only with German soldiers, moreover, they deal only with particular German soldiers in certain definite situations. Whether other soldiers of other races would react similarly in similar situations I do not know, but I believe not. For instance, the mentality of the American soldier differs from that of the German; and even in America, the northern soldier differs from the southern. A soldier from the city of New York has an entirely different viewpoint than the soldier who has lived as a farmer in the Middle West. He will, therefore, react differently and will require a different method of handling.

[10]

During the battle of Tannenberg, Hindenburg, Ludendorff and their staff were standing on a hill and observing a portion of the battlefield. While so engaged the well-known Colonel Hoffman, who was at that time G-3, came up to a young captain of the General Staff and said to him in a quiet tone.

"My friend, you seem to have nothing to do. Pay attention; in the village of X there is a Landsturm battalion. Call up its commander and say to him, 'A Russian cavalry brigade has made a deep penetration in the direction of the village of X. The Landsturm battalion is to counterattack and throw back the Russians.'"

On hearing this the young general staff officer became quite excited, and said, "Oh, Colonel, that Landsturm battalion consists only of old men over 45 years old. They cannot defeat a Russian cavalry brigade."

The Colonel answered, "Merely give him that order quietly, and if the battalion commander refuses to obey, ask him for his name. You will see that he will do it instantly."

The young captain gave the order over the telephone and the battalion commander, terribly excited, answered, "How can I attack a Russian cavalry brigade with my old men? That's impossible."

Then the captain said, "I have been directed, if such be the case, to merely ask you for your name."

"Oh, me," came the quick reply, "I did not mean it that way; certainly we will attack. I will have my unit forward at once, and in five minutes will be on the march. Your orders will be executed immediately."

And they were.

The fear of unpleasant consequences resulted in the disappearance of all of this commander's fears. With another battalion commander in different circumstances, the effect would probably have been entirely different. Colonel Hoffman had correctly estimated the probable reaction of this battalion commander.

A really classical example of this art of estimating a situation psychologically was shown in the year 1917 by a brigade commander. This General said, "Each of our three regimental commanders must be handled differently.

[11]

Colonel "A" does not want an order. He wants to do everything himself, and he always does well. Colonel "B" executes every order, but has no initiative. Colonel "C" opposes everything he is told to do and wants to do the contrary."

A few days later the troops confronted a well-intrenched enemy whose position would have to be attacked. The General issued the following individual orders:

To Colonel "A" (who wants to do everything himself):

"My dear Colonel 'A,' I think we will attack. Your regiment will have to carry the burden of the attack. I have, however, selected you for this reason. The boundaries of your regiment are so and so. Attack at X hour. I don't have to tell you anything more."

To Colonel "C" (who opposes everything):

"We have met a very strong enemy. I am afraid we will not be able to attack with the forces at our disposal."

"Oh, General, certainly we will attack. Just give my regiment the time of attack and you will see that we are successful," replied Colonel "C."

"Go, then, we will try it," said the General giving him the order for the attack, which he had prepared some time previously.

To Colonel "B" (who must always have detailed orders) the attack order was merely sent with additional details.

All three regiments attacked splendidly.

The General knew his subordinates; he knew that each one was different and had to be handled differently in order to achieve results. He had estimated the psychological situation correctly. It is comparatively easy to make a correct estimate if one knows the man concerned; but even then it is often difficult, because the man doesn't always remain the same. He is no machine; he may react one way to-day, another way tomorrow. Soldiers can be brave one day and afraid the next. Soldiers are not machines but human beings who must be led in war. Each one of them reacts differently, therefore each must be handled differently. Furthermore, each one reacts *differently* at *different* times, and must be handled *each time* according to his particular reaction. To sense this and to arrive at a correct psychological solution is part of the art of leadership.

Now with regard to other matters. We who have been in war know that the hardest thing we had to do was to lie quietly under hostile fire and wait for an attack. Why?

When a soldier lies under hostile fire and waits, he feels unable to protect himself; he has time; he thinks; he only waits for the shot that will hit him. He feels a certain inferiority to the enemy. He feels that he is alone and deserted.

I remember one day in 1916 in Russia. During the night we had relieved the Austrians. On the following morning the Russians began a heavy artillery preparation. We were unfamiliar with the terrain; we had no idea what troops were on our right and left; we did not know what artillery we had. I was alone with my company in the midst of an Austrian battalion. I didn't know my superiors. The Russians had been firing for hours, but our artillery didn't reply. I went constantly from dugout to dugout to see and speak to my men. They should at least see that they were not alone. Repeatedly they asked me: "Are we really entirely alone here; haven't we any artillery?" It continued this way for hours. Our telephone wires had been shot to pieces. Finally a tremendous noise came from our rear. Our own artillery was firing. At once high spirits returned. These soldiers no longer felt deserted. Each could see and hear that our side was now doing something. Each saw that he was being supported, and that everyone was ready to repulse the attack. In great defensive battles one will constantly hear the remark, when the enemy artillery is firing: "Where is our own artillery?"

It was the same with our aviators. If a hostile flyer was over us for merely ten minutes, the soldiers would begin to question: "Haven't we any flyers? Where are our flyers?" If our antiaircraft guns began to shoot at the hostile aviator, the soldier was at once satisfied; he saw that something was being done.

It is different during the attack. Here the soldier himself acts; he has something to do; he moves forward; he fires; he assaults and dictates the action of the enemy. At the moment of the attack he never asks "Where is our artillery?" From the beginning of the attack he feels him-

[13]

self the victor; he storms forward. He believes he can do everything by himself; he needs no support. As soon as the attack slows down, the cry for artillery is heard again.

In the Carpathian Mountains in February, 1917, my company was in position on top of a high mountain that dominated the terrain in all directions. In places the Roumanians were only twenty meters away. One day we were surprised by an enemy attack and pressed back to the edge of the mountain top. A hard hand-to-hand fight with bayonets and hand grenades ensued. At the end of an hour's fight we finally succeeded in pushing the Roumanians down the mountain. At the very beginning of the fight, I had seen the artillery observer, who had been in my trench, fall. From that time on I felt that my company and I were fighting alone without artillery support. In consequence, I called our regimental adjutant on the telephone and complained that the artillery had not helped us. The battery commander concerned came to me soon afterward and told me that his battery had fired about 300 rounds in my support during the fighting, that is, about five rounds every minute. I had not heard one of them. We had fought, acted, and in the excitement of the fighting I had not noticed that our artillery was firing.

This desire to act is, in my opinion, the reason why soldiers go so willingly on patrol. I repeat that it is extremely difficult to lie in hostile fire and wait, because one feels exposed to blind chance. On a patrol it is different. The soldier feels that his destiny rests in his own hands. He feels that he is not dependent on blind fate, that he is not forced to go this way or that, but can himself decide what to do. He feels that he controls the situation. For example, he may think: "That path over the hill seems dangerous to me; I don't know why, but I have that feeling most definitely; therefore, I prefer to go through the valley." He has the feeling that his action depends on his own will, and in consequence he can act in accordance with that will.

Here are two examples which show that this sense of security is a decisive factor. It is not a question whether security is actually existent.

In September, 1914, we were on a hill near Berry au Bac, near the Chemin des Dames. At our immediate right a road and a canal led down to Berry au Bac which was occupied by the French. A small stone house stood on this road. I had placed a picket of five or six men in this house to guard the road. One day I happened to be there when the French suddenly opened fire on the house with heavy artillery. A shell landed every minute. Everyone knows that single shells are far more unpleasant than a barrage, because there is time to wait and think things out. The first shell fell about fifty meters short; the second, about one hundred meters long; the third was again short; the next one hit close to the house. I noticed that my men were uneasy; they were now waiting for the shell that would fall in the middle of the house. I couldn't leave my men at this time, although my place was really not there. So we waited together. This waiting and this uncertainty made us nervous. We sat in the house and listened for every shell. We could tell exactly whether it was too short or too long, whether it would fall to our right or to our left. Finally the following thought came to me: "The walls of this house are very thick, in fact about a yard. If a shell bursts outside the house and we are in it, nothing can happen to us. If, however, a shell bursts in the house, then it would be better to be outside. Therefore, the best thing to do is to sit in the door and watch the shells. We can tell where they are going and we will be in a position to go either into the house or out of it." So I sat down on a chair in the door and was soon perfectly satisfied—so satisfied, in fact, that I went to sleep. This action on my part calmed my men to such an extent that they began to play cards. After a few hours the firing ceased.

You may laugh at my action in this case. I too am ready to laugh at it. My conviction at that time was nonsense. One cannot decide whether a shell will land three or four yards to the right or to the left. I have only mentioned the point to illustrate that it makes no difference whether or not the security is real; it is simply a question of feeling that it is.

Still another example. It was August, 1916. The great Russian offensive under General Brussilov had thrown the Austrians far to the rear. We were brought up by rail and then moved to the front in rapid marches to assist the

Austrians. For several days we bivouacked in a forest behind our artillery. Then, one night, we moved up close to the front as a reserve and our companies were scattered all over the terrain. As we were unfamiliar with the sector, an Austrian noncommissioned officer conducted our company to the front. We were halted under a large shed. We were happy to have a roof over our heads and slept soundly until morning.

When dawn broke I saw that this shed was entirely in the open and not more than 200 meters from an Austrian battery. This placed us in such a position that, if the Russians began firing at this battery, we would be right in the middle of their concentration. Furthermore, I could see a Russian balloon, therefore, we could not move out of our shed. My fears were soon confirmed; the Russians opened upon the Austrian battery with heavy artillery. One out of every three or four shots fell short bursting close to the shed in which my company was sheltered. Until night fell or the Russian balloon went down we could not move. The shells continued to fall around our shed. No one said a word. I noticed that my men were highly nervous. Several came to me and asked permission to go outside, giving more or less trival excuses. I refused, for it was apparent that they only wanted to reach a place of safety. The nervous excitement became tense. Suddenly a shell came down right in the middle of the company, but it failed to burst. Nerves were frayed almost to the breaking point. We were like a kettle which would soon boil over.

In order to obtain a feeling of security somebody had to act. Then I had a good thought; I called the company barber, sat down with my back to the front and told him to cut my hair. I must say, that in my whole life, no haircut has ever been so unpleasant. Every time a shell whistled over our heads, I jerked my head down and the barber would tear out a few hairs instead of cutting them. But the effect was splendid; the soldiers evidently felt that if the company commander could sit quietly and let his hair be cut that the situation was not so bad, and that they were probably safer than they thought. Conversation began; a few jokes were played; several men began to play cards;

someone began to sing; no one paid any more attention to the shells, even though two men were wounded a few minutes later by a shell which struck in the vicinity.

Two points stand out in this incident:

Instill a sense of security in the men; by so doing you will help them overcome their fears.

Do something to induce action among them. If they have been on the defensive for a long time, send out patrols even if there is no special reason for patrols. Patrolling instills a sense of self-confidence and superiority. I served for a long time under a regimental commander who demanded that one patrol be sent out from each company every night. Each patrol was required to bring back clear-cut evidence of its activity such as a prisoner or a piece of hostile wire. Soon there was a regular competition among the companies. Everyone wanted to go on patrol.

In the German army we use what we term "mission tactics"; orders are not written out in the minutest detail, a mission is merely given the commander. How it shall be carried out is his problem. This is done because the commander on the ground is the only one who can correctly judge existing conditions and take the proper action if a change occurs in the situation. There is also a strong psychological reason for these "mission tactics." The commander, who can make his own decisions within the limits of his mission, feels that he is responsible for what he does. Consequently, he will accomplish more because he will act in accordance with his own psychological individuality. Give this same independence to your platoon and squad leaders. It is certainly evident from training in peace that the more freedom allowed a subordinate leader in his training, the better the result will be. Why? Because he is made responsible for results and allowed to achieve them in his own way.

A few more examples at random will illustrate other aspects of this interesting subject of soldier psychology.

In August, 1914, we marched through Belgium towards Liege. It was a beautiful morning. The men sang. We were young, healthy, and we had the feeling of power and strength. On the road we saw the first dead. Singing ceased. Soldiers stared at their dead comrades. The seriousness of war suddenly appeared before their eyes. Perhaps they, too, would soon lie dead by the edge of the road. In absolute quiet, the company marched on. Then suddenly someone called to a dead man, "Seems to suit you to sleep; get up, it is breakfast time." All laughed. The seriousness of the moment had vanished in a joke. High spirits returned.

One day later. The first wounded soldier is seen. He is groaning in pain. A few comrades come up to him. One sees that the men are depressed. It is the first wounded man they have seen. Who will be next? Then someone says, "It is a pity that you are not dead; then I would have fallen heir to your beautiful boots." The wounded man stops groaning. He notices for the first time how fine it is that he is still alive.

It is the autumn of 1916. The division is involved in a terrible battle. The losses are heavy. At division headquarters one sees only worried faces. What will be the result of the battle? Suddenly a badly shattered and excited battalion commander rides up with his orderly. He states, "I desire to report that I am the only survivor of my battalion." A moment of tense silence follows. Then the division commander says: "You are in error, Major; your orderly is still with you." Immediately high spirits returned. These little examples show what miracles a word at the right time can work. It is a psychological reaction that defies explanation.

It is the autumn of 1914. We are just going into battle. I notice a man shaving himself. Yesterday before we went into battle I also noticed him shaving himself. I questioned him: "Why are you shaving?" He answers: "Since Liege I have shaved before every fight and I have never been wounded. At any rate it can't do any harm."

1917. The battle at Cambrai. A lieutenant with twenty men is defending a little clump of woods. He repulses several attacks. Another attack starts. Only a few Germans can continue the fire. They are out of ammunition. What shall be done? The lieutenant commands, "Fix bayonets. Attack. Hurrah!" The twenty men attack. Eighty English soldiers are taken prisoners.

Why do the English surrender? Why don't they merely laugh at the twenty Germans who are attacking?

February, 1917. It is in the same close combat on the mountain peak in the Carpathians previously described. Fighting has lasted an hour. We have not been able to drive the Roumanians back. In one place about six men are fighting. Their corporal is suddenly shot dead in their very midst. One of his men jumps up: "The Roumanians have killed our corporal," he yells, and charges into the midst of the enemy, knocking several of them down. The Roumanians run to the rear, the Germans after them. In five minutes we recapture the mountain peak.

These last two cases furnish examples of unexpected acts which, through their surprise effect, brought success. These things cannot be taught in peace.

We know that psychology is tremendously important in war. It is a field unlimited in extent, to which every conscientious soldier should give much time and study. Yet it cannot be *learned* as one learns mathematics. It must be *sensed*. Unfortunately we cannot formulate a set of psychological rules; human reactions can never be reduced to an exact science. War is governed by the uncertain and the unknown and the least known factor of all is the human element.

CHAPTER II

The First Days of War

Prior to the World War, little or no thought was given to acquainting officers and men with the impressions they would receive, and the perplexities and mental difficulties they would have to undergo at the beginning of operations in actual warfare. It is the purpose of this article to show how all ranks reacted under first battle conditions, and to suggest methods to eliminate, or at least to minimize, some of the mistakes made and to correct the erroneous ideas that are current in the first actve operations of a war.

It is important to bear in mind that at the outbreak of the World War Germany's troops were exceptionally well-trained, splendidly disciplined, and imbued with a remarkably high morale. All their leaders, both commissioned and noncommissioned, knew their duties thoroughly. The company of which we shall speak belonged to an active regiment. Its men were young and in excellent physical condition. But no one knew anything about the impressions of battle; no one knew how the human being would be likely to react to them; no one could say how he himself would behave in battle.

Under these circumstances it would have been helpful if the officers and men had known each other and served together for a long period of time. But this was not the case. For reasons that seemed sufficient at the time, almost all officers changed their duties on mobilization. It is true that they usually remained with the same regiment, but they took command of new units, where they were unknown to the men and the men unknown to them.

To-day this transfer of officers appears as a mistake. It is my strong belief that the heavy burden occasioned by the new impressions of battle would have been considerably lessened had there existed that feeling of unity and that mutual understanding which long service together engenders between officers and men.

One learns rapidly in war, and the first impressions, no matter how unimportant they may be, are lasting ones. Thus I find that the events and the impressions of those first days are still indelibly imprinted on my mind, although more than sixteen years have elapsed since their occurrence.

Our company crossed the border between Germany and Belgium on the 7th day of August, 1914. A short march after crossing the border brought us to a little village. The sun was just about to go down when an airplane suddenly appeared above the column. We had never marched into a hostile country before and we had never met an enemy. Everyone was nervous and excited over the events which they felt were sure to come. For this reason they were at once certain that an airplane in hostile territory could only belong to the enemy. Suddenly a shot was heard; then several; and in a few seconds the entire company was shooting. Next, machine guns could be heard somewhere in the distance taking up the firing; then artillery shells were seen bursting in the air. Now every doubt disappeared. If the artillery was firing, the aviator must be an enemy. Even the driver of the field kitchen shot at the poor aviator with his pistol. The aviator continued to fly for some time and then seemed to sink to earth. The psychological excitement was satisfied by the wild shooting. The men raised a loud cry of joy because they believed they had hit him, and nearly the entire company started off to take the aviator prisoner. One by one the men returned with foolish expressions on their faces, like a young hunting dog which has been vainly chasing a rabbit. Next day a division order was received which began as follows: "It is only due to the very bad shooting of the troops that today two German aviators are still alive . . ."

In the meantime it had become dark and the question arose as to where the company would sleep. Thirty to forty houses were available in the village for billets. It would have been natural and proper to have used most of them. But remember the mental condition that existed. It is true that no enemy had been heard or seen. However, since they were in a hostile country, the men believed that the enemy could appear at any moment. As night closed in it brought with it the ancient human urge for companionship and for the sense of security afforded in numbers. The outcome was that the entire company billeted together in a single large house, with the natural result that no one got much rest.

Now came the question of security. We had been taught in peace that in open warfare the enemy can only approach at night over unknown terrain by marching on roads. Therefore, under such conditions as ours, it was only necessary to place sentry posts on the main streets. But all this was forgotten. One platoon, about eighty men strong, was given the mission of providing the security. The young platoon leader was just as excited as everyone else. He promptly put his entire platoon on guard. This meant that one-third of his men, thirty in all, were simultaneously on sentry duty, the others were used as reliefs. Sentry posts were established all around the house. Having established them, the young officer went back into the house.

Suddenly a shot rang out, then another, then several in concert. Then a message came in, "The Belgians are coming." The platoon leader ran outside to a sentry and asked him, "What's wrong?" "There they are behind that bush," said the sentinel as he fired. Nothing moved. Now a shot rang out in another place, then at another. Nothing could be seen. The officer ran from one sentry to another seeking to quiet them. For a while everything would be still, but in a few minutes the firing would break out again. As soon as one sentry fired, several others did the same thing.

I often experienced such panic firing later on in the war. Generally there was no reason for it. An especially good example occurred in September, 1916. During the great Russian offensive under General Brussiloff, I relieved a company of Austrians with my own company. The commander of the Austrian company told me that heavy firing took place every night. It was true. Night had scarcely fallen when the Russians began firing. My company consisted solely of experienced men who had already been in a large number of fights. Not one of my men fired be-

[22]

cause they could see nothing to fire at. They waited impatiently for the Russians to attack. Nothing happened. The next night the same thing occurred. On the third night the firing was appreciably less, and a few days later there was no firing at all during the night. The reason for this is always the same; the soldier's psychological excitement is satisfied by firing. The result is that men use up their nerve strength and throw away ammunition.

So it was our first night in Belgium. Finally the night came to an end and dawn began to break. The platoon leader ascertained that hundreds of shots had been fired. The men searched the immediate vicinity confident that so much shooting must have caused hostile casualties. They searched and searched and finally found—one dead cow.

The foregoing event happened with exceptionally well trained troops. The reason must be sought in psychology. It was the first incursion in hostile territory, and the experience was new to the men; it affected them not only in their capactiy as soldiers, but as human beings.

On the next day the march was continued. The enemy not having been encountered the day before, the men reasoned that they would certainly meet him to-day. The psychological excitement diminishes only after one has passed through his first fight. The company marched at the head of the battalion. A young officer was ordered to take a few of his men, form a patrol and move on ahead toward a village which lay five kilometers to the front. He received no information of the enemy because the company commander himself did not have any.

As he approached the village there were no signs of life. It appeared to be dead and deserted. A hot sun shone down from a blue sky; there was not a breath of wind; all was quiet, mysteriously quiet. What could be in the village? It seemed certain that the enemy would allow the Germans to come close and then suddenly open fire on them. The young officer therefore left the road and approached the village through a little gulley in order to reach this mysterious place from the flank. He frequently stopped and looked through his field glasses. On one occasion he thought he saw a movement in the village, but then it again became still. Now he came to the gardens behind the village. Every moment he expected to meet the enemy. Soon he would be able to see the main street. As he wanted to observe the enemy, he began to creep and crawl. Carefully and slowly he worked his way through the garden to the road.

There stood—a battalion of a German regiment, which was a part of his own brigade. This battalion had spent the night in the village. Later he learned that his battalion was almost at the tail end of the division and that everywhere German troops were moving forward.

During experiences such as these one learns very quickly. The members of the company soon believed themselves to be seasoned soldiers. As a matter of fact, years are required for the process. It is still better to say that there is no end to learning, especially in war.

Next day brought a long, tiring march toward Liege. It was almost dark when the company halted. The men were billeted in a chateau. After a few hours of rest they were awakened. It was still dark. The company fell in, joined the other companies of the battalion, and all moved forward at a rapid pace. The men were told that that day they should storm one of the forts of Liege. Suddenly they heard a tremendous noise which was followed by a loud whistle over their heads. Rapidly this whistling came closer; it seemed about to come down on them. "That's heavy artillery," must have been the thought in most of their minds, because a large percentage of the company dropped to the ground. Nothing happened, however. It was their own artillery that had opened fire and the first real shell of heavy artillery that they had heard in the war. Although it had given them a terrible fright, the men were soon laughing again and those who had laid down on the ground showed their shame.

At this time they were not experienced enough to decide whether it was their artillery or that of the enemy; whether the shell came from the front or the rear. Later they learned all about shells; they could determine, not only whether it was their own or the enemy's, but they could distinguish the calibre of the shell and the type of gun or howitzer that it was fired from. They soon learned when it was necessary to take cover. After long experience, they could tell explosive from gas shell. Those who served on different parts of the front were able to judge by the sound and intervals between shells, whether the artillery that was firing was English, French, or Russian.

Toward evening of this day the fort was stormed.

These matters that have just been related must seem trivial. However, long experience has taught us that we know too little of the small incidents of war, especially of combat at the outbreak of hostilities. In peace we learn how to lead companies, battalions, regiments, even divisions and armies. We learn in books and by maps how one fights and wins battles, but we are not instructed in the thoughts, the hopes, the fears that run riot in the mind of the front-line soldier. We are not told how we can help him in his mental battles.

This problem will be even more serious in future wars, when machinery rules the battlefield, than it was in the past. In peace maneuvers such matters cannot be depicted. We can learn only from experience or by analogy from a searching study of military history. As leaders we must constantly seek some means to prepare our soldiers for these grave psychological blows that war strikes at morale and nerves.

The three important lessons to be learned from these little stories are:

(1) At the commencement of war, soldiers of all grades are subject to a terrific nervous strain. Dangers are seen on every hand. Imagination runs riot. Therefore, teach your soldiers in peace what they may expect in war, for an event foreseen and prepared for will have little if any harmful effect.

(2) As leaders be careful both in sending and in receiving reports. At the commencement of a war ninety percent of all reports are false or exaggerated. Learn in peace as you prepare your map problems, field exercises, and war games, to give false or exaggerated reports; otherwise, your subordinates will become accustomed to accepting all information they receive at its face value.

A short story from my own experience will illustrate this point:

During the afternoon of the last day of which I spoke,

[25]

I was sent on patrol against the fort. I finally reached a hill, and in order to see better, climbed a tree. My men remained below on the ground. Suddenly there was a loud crash, and an artillery shell burst, just in front of us, or in the branches of my tree. It was the first hostile artillery shell I had ever heard. As a result, I fell ignominiously off my branch to the ground. My first thought was, "Now I am dead." As I hit the ground, however, my bones told me that I was entirely alive. My men had run away. I learned later that they had run all the way back to the regiment, and to explain their desertion of an officer, had stated that I had been killed. This report was believed.

(3) You saw how the company on the first day of the war marched into hostile territory without any knowledge of the situation. You must learn in peace that in war of movement you must go ahead without sufficient information of the enemy. Otherwise, you will send an equally stupid lieutenant on patrol, who will crawl all over the landscape for hours and finally locate your own troops.

CHAPTER III

The 14th German Division in the First Battle of the Marne

After a rapid advance through Belgium and north France the German armies, at the beginning of September, 1914, were stretched on a long arc from Paris to Verdun. The last days had seen much hard marching. The German soldiers were still marching from dawn to dark; the packs were heavy, boots hurt, but they all knew the old principle of the German army: "Battles will be won by legs." Those in the front line felt that the great decisive battle was still to come.

The commander-in-chief believed it impossible to pass the gigantic fortress of Paris without security against it. Therefore, on September 5, he ordered the First and Second Armies to turn to the right against Paris in order to secure the remaining armies. That order found the First and Second Armies in the situation shown on Sketch 2 (at end of book). The movements directed by this order, coupled with the decision of the French to stop their retreat and attack the German right flank, brought on the battle of the Marne.

I shall first describe the events as seen by the soldiers at the front; then I shall discuss the reasons that led to these events.

Let us follow a young officer who served with one of the infantry regiments of the 14th Division, and the battalion to which he was assigned. (See Sketch No. 1 at end of book).

On September 5, the 14th Division, on the right wing of the Second Army, made a long march, passing east of Montmirail. The troops of this unit crossed the Petit Morin River and spent the night in a small village south of the river. At daylight September 6 there was a loud booming of artillery in the south. Instead of marching toward the sound of the guns, however, the division marched to the north. The men were uneasy, because they were unable to learn why they had to march to the rear. Heretofore they had always marched forward. They had only been at war for one month and had yet to learn its ups and downs.

About noon, it was learned that the division was to be in army reserve. A little later a halt was made on both sides of the main road from Chateau Thierry to Fontenelle. Good spirits soon revived. The men were contented to stay there all afternoon, because they knew war well enough to know that when the reserves are allowed to rest the battle must be going very well. At nightfall the sound of artillery finally subsided.

Early on the morning of September 7, the division made a short march into a wood south of Artonges. Here the men saw something which was far from reassuring—many badly damaged German supply wagons, returning hastily from the front. The drivers told of a German retreat, of heavy casualties, of a defeat. The men grew anxious about this battle, that they could hear, but not see.

About 8:00 AM the battalion, to which our young officer belonged, marched to a position southwest of Villemoyenne and began digging in. The men could not understand the situation. To date they had been driving the French to the south. Yesterday they had marched to the north. This morning it was rumored that their troops had been forced to retreat. Now they were digging in, with their front to the west—that is, toward the outside flank. Where could the First Army be? Everything seemed to indicate that the French would attack the right flank of the Second Army. The officers felt this anxiety among their troops. However, there was not much time to think about it, for at 11:00 AM orders came to stop digging and start on a forced march to the east.

Their destination was Fromentieres, located behind the center of the Second Army. The march was difficult, hot and dusty, and to complicate matters still further the battalion had to cross several long columns of ammunition and supply wagons. About 1:00 PM it halted, though it had not yet reached Fromentieres. Then it turned about and marched back to the vicinity of Artonges.

These frequent changes made an unfavorable impression on the troops. Apparently the high commanders didn't know what to do. Since the troops had had but little experience at war, all these seemingly confused and vacillating orders struck a serious blow at their morale.

About 5:00 PM the battalion again reached the neighborhood of Artonges. At 8:00 o'clock in the evening, while the officers were still discussing the events of the day and trying to discover the reasons behind them, they received a new shock. The battalion commander told them that the French had penetrated the left wing of the army, and that they were to march to the assistance of the left wing immediately—without rest and without regard for march casualties.

It was pitch black when the battalion started on this forced night march to the east. Part of the time the march was made crosscountry. For five minutes out of each thirty minutes the troops ran in double time. They believed that the battle must be going badly, otherwise such a march would not be necessary.

About 1:00 AM, the battalion reached its goal near Champaubert. The march casualties had not been very heavy. One company, for example, lost only 10 men out of 130. In war, march casualties are never heavy in night marches or in a retreat. The reason is simple; no one likes to leave his unit when he doesn't know where he is, or when he is fearful of capture.

At Champaubert the soldiers dropped to the ground and fell asleep at once. After three short hours of rest they were aroused. They had expected to attack at daylight but instead they again marched to the east. Nobody knew why.

Now we shall briefly consider the reasons for these movements:

On September 5, the left wing of the First Army was situated farther south than the Second Army. (Sketch 2.) If the turning movement ordered by the commander-in-chief were to be made, the Seventh Corps, with its 13th and 14th Divisions, was superfluous at the front at that time. Therefore, it was designated as the army reserve and was moved north of Montmirail in order to clear the pivot of the army. Meanwhile, the other corps had to wheel to the right in a large circle.

On September 6, however, all corps marching to the south were heavily engaged and the 13th Division had to help the hard fighting IX Corps. Thus only the 14th Division remained in army reserve. On the same day it became evident that the IV Reserve Corps could not secure the flank of the army on the Ourcq River by itself. Therefore, the First Army was forced to order one corps after another from the south to the Ourcq River, finally even the III and IX Corps. Thus the rumor of a retreat started in the 14th Division.

The gap that resulted from these movements was secured only by some cavalry divisions. This situation made it necessary for the right wing of the Second Army to fall back, the attack to gain a decision being continued by the left wing alone. Thereupon it fell to the lot of the 14th Division, located behind the right wing, to secure the right flank of the army. This explains why it was ordered to dig in with its front to the west. Meanwhile, a bloody struggle began along the front of the entire army. Naturally the army commander desired his reserve behind the center of his front in order to be able to engage it wherever he might need it. As the reports from the gap between the First and Second Army did not indicate that the situation there was critical, he ordered this movement. Hence the march to Fromentieres. This movement was scarcely completed, however, when the army received fresh information indicating that immediate security for the right flank was essential. Consequently, the 14th Division received an order to march back behind the right wing of the army. Later, on the evening of September 7, still other information arrived at headquarters of the army to the effect that the French had made a penetration between the X Corps and the Guard Corps. The only available reserve, the 14th Division, had to make the difficult night march.

The situation at this time (evening, September 7) was as follows: the right wing of the Second Army had retreated, but the French and the English had followed very slowly. On the left wing, however, there seemed to be a weak point in the hostile front. It might be possible to penetrate the French lines at this point, but if this were to be done, it would have to be done promptly, as the situation on the right wing could change very quickly. By this time it had become clear that there was no immediate need of the 14th Division where it was. Therefore, as it was the only available reserve, it was aroused in the early morning of September 8 to help win the battle.

Let us consider these experiences of the 14th Division a bit further.

On September 5 the German leaders believed they were confronted by an enemy, who was continually retreating, and they based their decisions on this assumption. The Germans were therefore surprised on September 6, to find that the French were no longer retreating and were even launching counterattacks. Such a lack of knowledge of the enemy's intentions and actions may always be expected in open warfare. In open warfare, we will never know exactly where the enemy is, how strong he is, or what he intends to do. War is not as easy as a map problem. Leaders must nearly always issue orders without exact information. Our mission and our will are often the only things untouched by obscurity. These will frequently form our only basis for an order. If a leader awaits complete information before issuing an order, he will never issue one.

We saw that the Second Army received an order on September 5 to furnish security against Paris. This order had to be given immediately, although general headquarters had but meager information. This decision proved to be correct. Two days later the 14th Division was ordered to make a difficult night march from the right wing to the left wing; in this case the decision of the army was based on false information. The 14th Division was involved in an unnecessary effort. In war much information is false or misleading; therefore, we must carefully weigh all reports before accepting them.

Owing to the uncertainty of battle and the lack of enemy information, it is essential that a leader keep out reserves. Reserves, however, are meant to be used. In battle it is only the fighting soldier that brings an advantage.

As the information of the enemy changes, it is the tendency of a leader to move his reserves according to his momentary conception of the situation. In the foregoing example, the 14th Division, which was the Army reserve, was so moved. Its different movements, some of which were necessary and some not, illustrate the fact that in war there will be much marching and countermarching, the reasons for which the troops will not perceive. They must be taught in peace to expect this. The leader, of course, should always have regard for the fact that useless marches not only diminish the physical capacity of his troops, but seriously affect their morale as well. On the other hand, soldiers should be made to realize that the changing conditions of battle will frequently result in marches which later prove to be unnecessary. They must be taught to endure this with calmness and fortitude.

Let me repeat the two outstanding lessons that should be learned from the foregoing account:

1. In open warfare a leader will have to give his orders without having complete information. At times only his own will is clear. If he waits for complete information before acting he will never make a decision.

2. The strength and morale of reserves must be conserved in order to reap the full benefit of their freshness when they are finally committed to action.

The second part of this narrative deals with the attack made by one battalion of the 14th Division. The 14th Division had a difficult task in this attack, which was to decide the outcome of the battle. It had to force the crossing of the broad swamp of the Petit Morin River south of the village of Joches. (See Sketch No. 3 at end of book).

The terrain was as unfavorable as possible. The swamp, impassable even to foot troops, was bridged by a single road. The French, located south of the swamp, could fire with artillery and machine guns on both Joches and this road. A long row of high trees beyond the swamp prevented the German artillery from observing Hill 154.

Marching from the north, the battalion with which we are dealing reached the north edge of the village of Coizard at about 8:00 AM September 8 and prepared for action in the gardens of that village. The day was boiling hot. A blazing sun glittering in the blue sky dazzled the eyes. The young officer, who commanded the 2d Company, moved to the south edge of the village in order to reconnoiter the terrain to his front. From here he could see Joches, the formidable swamp, and beyond the swamp Hill 154 interlaced with hedges and dotted with sheaves of grain. He could recognize some Frenchmen upon Hill 154, as their red pantaloons were clearly visible in the morning sun. There was no firing. Behind him he saw some German batteries, just reaching their positions, and he heard his soldiers singing a song from home.

About 9:00 A. M. the first scouts left Joches and hastened forward to cross the swamp. The French immediately opened fire on the road and the village with artillery, machine guns, and rifles. The young officer, still observing, saw the scouts jump into the ditch east of the road and hurry on. At this point he was called back to his company where the battalion commander gave him the following order:

"One battalion of another regiment starts the crossing of the Petit Morin River. This regiment will follow, the 2nd Company at the head. The objective is the village of Broussy le Petit."

That was all.

The company officers and men knew neither the location of the enemy nor his strength. They had not been told whether or not they would be supported by artillery. They did not know who was on their right or left. They only knew that they had to attack and that they would meet the French beyond the swamp. This was always our experience in open warfare whether on the Western Front, or the Eastern Front, in Rumania, or in the Caucasus; when an attack started we never had more than poor information of the enemy.

Before the attack was launched, the company commander told his men to fire calmly and with the correct elevation. He told them this because in earlier fighting he had observed that many excited soldiers forget to take the proper sight setting. The company then moved forward and soon reached the village of Joches.

The French artillery fired on the village but the company

passed through it on the run without any casualties. Upon reaching the southern edge of the village, it was seen that the last elements of the leading battalion had not yet crossed the swamp. The company commander therefore moved his company from the road to cover behind a nearby building. From here he watched the rear squads of the battalion ahead run forward in the ditch beside the road. Several soldiers were hit while crossing the bridge. The French artillery fired like the devil on the road and on the village of Joches, which soon began to burn. The company was bunched and lying behind its commander. No one talked. Everyone looked at him. Suddenly, with a terrible crash, a shell hit in the middle of the company, killing or wounding ten men. All the men sprang up; there was a big tumult. Suddenly there, in the middle of the company, was the Brigadier riding his horse. He was calmly smoking a cigar as if nothing had happened.

He commanded: "2nd Company, fall in. I hope I won't see such a picture again. If I do, I shall confine some of you, when we return to our garrison."

Such extraordinary coolness had a marvelous effect on the troops. The company fell in promptly, formed into platoons, and moved forward as quickly as possible in the ditch beside the road. By 10:00 A. M. it had crossed the swamp and the bridge with only three other casualties.

On the other side of the swamp the company commander assembled each of his platoon separately. Dead and wounded from other units were sprawled on the ground near the long row of trees. The company commander had no idea where the enemy was or for that matter the location of his own supporting troops. He could only hear firing beyond the trees. Shells and machine-gun and rifle bullets whizzed over constantly.

The company advanced with two platoons, while the third took cover among the trees. As soon as the forward elements left the cover of the trees they came under a heavy fire. The company commander, seeing German soldiers firing about 100 meters before him, rushed his platoon about 50 meters beyond their lines and opened fire. Here the advance was stopped by machine-gun fire from "la Verrerie Ferme" to the left flank. As this fire was causing heavy casualties the company commander sent a runner to his third platoon with an order to storm the farm. After half an hour this disagreeable flank fire stopped. On the following day it was learned that the third platoon had taken the farm with only light casualties.

Meanwhile the company was involved in a fire fight but nothing could be seen of the French lines except a little smoke coming out of the hedges and grain sheaves.

The company commander, in the midst of his two assault platoons, noticed that soldiers from other companies were intermingled with his platoons. About this time the firing line was reinforced from the rear and the soldiers of the company we are following became intermingled with companies of two other regiments. All were lying in one line. The row of trees, a little to the rear, seemed to be the assembling point for all troops who crossed the swamp. By the end of half an hour all units were hopelessly mixed. There was no leading; the officers were all in the front line and could only see a little distance to the right and left. They were not sufficiently experienced to know that the place of the commanders of companies and higher units is not the first line.

Thus it came about that three intermingled battalions were engaged in battle on a front of 700 meters. The artillery, handicapped by poor observaton, rendered ineffective support. Despite constant efforts to advance, the line moved forward no more than 100 meters. About 1:00 P. M. the last battalion of the regiment advanced on the right of the road, but it, too, was stopped by French artillery and rifle fire.

Although the men fired calmly a certain tenseness was evident owing to the fact they were unable to advance.

Suddenly, about 2:30 P. M., a battery of German light field howitzers came galloping out of the burning village of Joches and approached the bridge. The French artillery fired like hell, but the battery passed the bridge safely and went into position 300 meters behind the infantry. From this position, it opened a prompt and effective fire. At the same time some companies of another infantry regiment arrived on the line. An officer, feeling that this was the time for the decision, shouted to his bugler lying beside him:

"Blow the signal for the attack." The bugler calmly stood up and blew the signal that was well known to the soldiers. Other buglers repeated the call. All along the line, the men sprang up, fixed bayonets, and with a great shout rushed forward. The French, however, were brave men and stood their ground. Their fire became still heavier, but the critical period of the Germans had passed. Now that they were attacking, they were filled with enthusiasm and with a determination not to be stopped. The attack was successful. Some of the French escaped before the attackers reached their lines; the rest fell, fairly fighting, or were taken prisoner.

From the top of Hill 154 which they had won, the Germans looked down on the village of Broussy le Petit toward which the French were retreating. Everyone thought: "We have won, the battle is over. Now we start the pursuit, in order to keep the French on the run." Without waiting for orders, they rushed downhill, to capture the village of Broussy le Petit.

Suddenly they were struck by terrible machine-gun fire from the valley to their front and, at the same time, the French artillery opened fire upon them with all its force. In a few minutes the Germans had suffered enormous casualties and were forced to halt and lie down. They tried to take cover, but this was impossible as they were on the hillside sloping down to the enemy. They were caught in a hailstorm of fire. Their own artillery was unable to assist as it had no observation of the valley in front of the infantry.

At this tme, our young officer found himself surrounded by men from a dozen different companies. No one was of his own company. To add to the difficulties of this critical moment, the German medium artillery suddenly began dropping shells in the midst of their troops. That was too much. At first only a few soldiers, but soon more and more, went back to the other side of the hill in order to take cover. The officers and some of the non-commissioned officers sprang up to keep the men from going back but they were only partially successful in stemming the movement to the rear.

The reason for this retreat is to be found in psychology. The men had undergone a forced night march and a heavy struggle throughout the morning. After their successful attack, they though: "The battle is over; we have only to pursue; our work is done". Then came this terrible fire and heavy casualties. Finally shells from their own artillery fell among them. The accumulation of diverse and unexpected events proved too hard a test for these soldiers, still young in warfare and untrained in the experiences of battle.

More and more soldiers went back. It began to look as if the whole line would break when suddenly, six horse-drawn guns galloped forward over the ridge, moved right into the front lines and went into action. This gave new courage. At the same time, fresh companies advanced and once more the bugles blew the attack. That was the right moment. Every man sprang up and charged toward the enemy who was only four hundred yards away. After some sharp hand-to-hand fighting the French retired across the swampy ground to the village of Broussy le Petit. By this time, however, night had fallen and the village could not be taken. The Germans spent the night in the lines they had reached but there was little rest. During the night units were reorganized, field kitchens brought up warm meals, and the wounded were sent to the rear.

On September 9, about 4:00 A. M.—it was still dark—scouts sent back word that there were no French in Broussy le Petit. Our company commander aroused his company at once, sent the message back to the battalion commander and started forward for the village. Entering it without opposition, he advanced to the hills beyond. Even from the top of these hills he could not see any enemy, but at this point he received an order from his battalion commander to halt the advance and to await further instructions. About 1:00 P. M. he received an order to retire behind the swamp. Neither the commander nor his men could understand this as they felt that they had won the battle on the day before.

They withdrew without receiving any fire from the enemy, crossed the swamp and dug in beyond. On the following night they started the retreat. The battle of the Marne was over.

ANALYSIS

It is not my task to discuss the reasons for this retreat. I shall concern myself only with a consideration of the events of the action and the lessons to be drawn therefrom.

The casualties on September 8 had been very heavy. The company of our young officer lost 62 of the 120 men with which it entered the engagement, more than 50 per cent.

In this struggle most of the factors typical of the first months of a war are in evidence. Let us analyze a few.

1. You will recall the poor attack order this company received. This was the only order it received during the entire day. Although our leaders were long in service and well trained in issuing orders, we received only one order and that was a poor one. The reason for this becomes clear when it is remembered that our field exercises and war games in peace were based on good information of the enemy. But here there was no information and the leaders did not know what to do. It is even probable that they thought the enemy was not present in strength, because there was no information of him. There is a valuable lesson in this. Our field exercises, our map problems and our war games should be conducted with as little information of the enemy as possible. Our tactical problems in peace will then more nearly conform to the realities of war.

2. It will be remembered that at the very outset of the attack, all units became intermingled. What was the reason? Each unit, immediately after crossing the swamp near Joches, hastened forward without waiting for the other troops. The result was a convergence on a narrow front and much confusion. Company after company, and battalion after battalion moved forward to the attack and was stopped. The lesson from this is, prepare your attack well. The time devoted to such preparation is not lost. In this case, after crossing the only available bridge, the various units should have taken cover and waited until the advance could be coordinated and launched on a broad front.

3. The men, although well trained and of a high morale, were inexperienced in war and reacted strongly to early impressions. The reason may well be that they had not been psychologically prepared for the severe trials they were called upon to undergo. The conclusion to be drawn from this is obvious: we must teach our men in peace that battles differ greatly from maneuvers and that there will often be critical

periods when everything seems to be going wrong. It is exceedingly difficult to teach men what to expect in war, but something along this line may be accomplished if we study military history and teach its lessons to our soldiers.

4. Officers must lead their troops personally and set them an example. On September 8 it was a battery, led by officers and boldly advancing into position, that was responsible for the success of the action. Other phases of the battle similarly found officers leading the advance. This does not mean that it is either proper or advantageous for all officers to be in the front line where observation and knowledge of the battle are limited to a few meters right and left. There will be many occasions, however, when officers must set an example in order to inspire their troops to advance or to hold their ground in the face of almost certain death.

CONCLUSION

In conclusion I wish to emphasize the following lessons:

(a) The strength of reserves should be conserved for the hour of decision.

(b) Officers and soldiers should be prepared in peace for the psychological impressions of war by studying and teaching military history.

(c) Troops must have personal leadership and example from their officers in battle, but all officers must not be so involved in the front line that their efforts are localized and that they are unable to handle their units as a whole.

(d) Plans and dispositions for an attack must be made carefully. A modern attack requires ample time for preparation.

(e) When the situation demands it decisions must be made promptly without waiting for good information. Field exercises, map problems and war games should be based on poor information of the enemy. If real war brings us better information we will have an easy time.

CHAPTER IV

Leavening Raw Troops with Battle-tried Veterans

I have already spoken of the lessons the German Army learned during the early days of the war on the Western Front with well trained, but inexperienced troops. Let us now consider a further experience with young troops, this time, however, under entirely different conditions.

Shortly after the outbreak of hostilities several new corps were trained in Germany and later in the autumn, sent to Flanders, to penetrate the hostile line. These corps were formed from volunteers, generally students from the universities. Entire classes stood beside their professors in the ranks. Their equals in enthusiasm and morale will not be found very soon. Their training, however, was insufficient. It had been conducted by officers who had been on the retired list for years or by reserve officers who were too old to fight with the active regiments. In October and November, 1914, these old officers led the new corps into battle in Flanders where they met an active enemy who by this time had several months experience in war. The new German units attacked with tremendous drive and the men fought like heroes but their every effort failed with unbelievable high losses.

The question now arose of placing still other corps in the field in the shortest possible time. The mistakes that had been made with the October corps would not be repeated with these new units. In consequence, about the first of January, 1915, officers, non-commissioned officers and men who had already had considerable combat experience were transferred to these new corps. As a general rule these officers had been wounded and were now being returned to the front from the hospitals. Every officer in these fresh units had had some experience in the war. Many of the non-commissioned officers and men were similarly experienced. The composition of the new units was such that all the officers, about one-third of the non-commissioned officers, and from one-fifth to one-sixth of the men were experienced. These troops were trained as units for only three weeks. They were then assembled into divisions and given two training exercises as divisions. They were then transported to East Prussia to take part in the winter battle of the Mazurian Lakes.

About the 10th of February one of these young regiments detrained to the north of Insterburg and marched into a little village which was called Krauleidschen. Here it remained two days during which the men were allowed to rest. It was terribly cold. The snow was about three feet deep. In every squad there were one or two men who had already been to the front. The others were young men, twenty to twenty-two years old. In every house there were some of these experienced soldiers. The young soldiers constantly gathered about these veterans, listening earnestly to their stories. These stories, naturally, had to do with the war, what it was like and what the old soldiers had learned. They were proud now to be old experienced veterans and they were especially proud that they could teach these greenhorns, who knew nothing about war.

They kept saying: "We want to tell you about real war. Listen to us. We have seen things entirely different from this." And then they would begin a new story. Or, they said: "We are fighting now under Hindenburg who has already beaten the Russians so many times. If Hindenburg is commanding us we are sure to win. Just you wait and see; we shall capture thousands of Russians."

This confidence in their commander was shared by the officers also. Everyone was inspired by the fact that they were fighting under Hindenburg. His fame at this time was already secure. He had won the battle of Tannenberg, the modern Cannae, had defeated the Russians at the Mazurian Lakes, and had conducted those two splendid campaigns in southern Poland and at Lodz. The officers looked into the future with the utmost assurance.

After a delightful rest the march was resumed to the front. The march was cold, long and terribly difficult through the deep snow. The columns stopped often and

[41]

then after short checks, marched on. But every now and then the thought arose, "Hindenburg is leading us; that's all that's necessary." During the night the regiment arrived just behind the front. There were a few unwired trenches, garrisoned by a handful of old Landsturm men. The Russians were at the edge of a wood about 400 yards away. The whole front was quiet. Only a few artillery shells burst in the lines during the night.

It was still dark when the Germans went into the trenches. In a surprisingly short time, the companies were formed for the attack and each man took his proper position in line. No questions were asked and there was no mix-up. Everywhere one saw the old veterans at work, showing their young comrades how to behave and telling them what they should do. They knew exactly what would happen. Every now and then one would say, "Listen to me and you won't have any trouble." This calmed the young inexperienced men. They no longer felt any worry for the future.

When day broke, the Germans attacked. The Russians were completely surprised. Their infantry fire was weak. Their artillery scarcely fired at all. In a few minutes the Germans were in their trenches. It was a pleasure to see how the young recruits attacked. They watched their old experienced comrades carefully and did what they indicated. It was a short action and the Germans were victorious. Immediately the pursuit was taken up. The younger soldiers now trusted their seasoned comrades more than ever. These in turn trusted their commanders. And finally everyone said, "Yes, we are fighting under Hindenburg."

Except for scattered detachments there was no longer any enemy in our front. Once in a while a few Russians were encountered but their opposition was quickly broken. The march was continued the entire day through heavily wooded terrain. It was bitterly cold. Although marching on a good road, the snow was so deep that the men at the head of the column had to be relieved every thirty minutes. On they marched throughout the day and night, not to the east but to the south. Despite the heavy pack and the difficult going, morale was good for these soldiers had been victorious. They believed that this long hard march had been ordered so

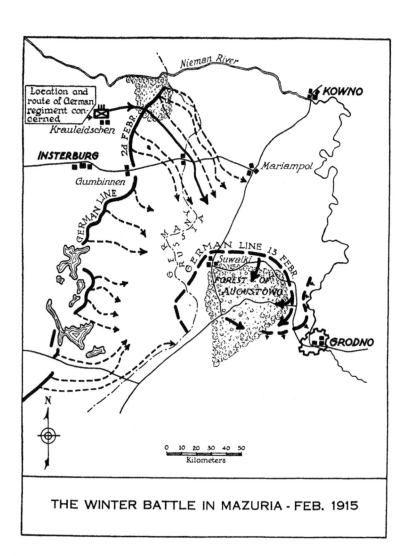

THE WINTER BATTLE IN MAZURIA - FEB. 1915

that they might encircle the Russians. Was that not the way Hinderburg had handled the enemy at Tannenberg?

As a matter of fact, their estimate was altogether correct. The army had succeeded in penetrating the hostile south flank. A tactical victory had been won. It was now a question of changing it to a strategical victory. In accordance with the original plans the victorious wings now sought to encircle the Russian center which had been held in position by repeated attacks. To accomplish this it was necessary for both wings to make tremendous marches.

Finally they approached a fair sized town. They were already rejoicing at the prospect of fine billets when the column suddenly bent from the main road and moved to the east. When some of the young recruits began to grumble, the old experienced soldiers growled at them: "You dumb recruits, are you cleverer than Hindenburg? What do you know about war? If we are satisfied, you ought to be. We were making marches when you were still at home at your mother's apron strings."

Morning came, but the march continued. A heavy fog hugged the ground. It was impossible to see more than 100 yards in any direction. Suddenly the column halted. The company commanders of one battalion were assembled and the battalion commander issued the following order: "About two kilometers in front of us is the main road to Russia. It is possible that we will find the enemy on this road trying to escape to the east. The battalion will deploy, advance toward this road and gain possession of it. The 3rd and 4th Companies leading, will move forward on both sides of the road on which we are now marching. The 1st and 2nd Companies will follow on this road at 500 meters. I will be at the head of the 1st Company.

The order contained no information of the enemy, because there *was* no information. Yet, in spite of this, a decision had to be made. The general situation demanded it. An order had to be issued.

The leading companies began their advance each with one platoon in the front line and two in reserve. I believe it was correct to place but one platoon in the forward line because no one knew what would happen in the next few minutes. Scouts went forward, each platoon sending out two pairs. Each pair of scouts consisted of one experienced soldier and one recruit.

In this manner the companies approached the highway. Suddenly one of the young recruits came running back. He reported: "Three hundred meters in front of us is the road. Russians are on this road marching towards the east." At the same moment the battalion commander came forward on his horse. He merely ordered, "Attack at once".

The companies continued their forward movement. Suddenly they saw the highway directly in front of them crowded with trains and artillery marching eastward. There was a yell, a few scattered shots, and the Germans charged. The next moment, they were on the highway in the middle of the Russian artillery and trains. There was terrible confusion. The Russians were so completely surprised that the thought of resistance never entered their minds. A few horses were killed and a few Russians escaped in the fog. The rest were taken prisoners with all their guns.

On the next day the division attempted to march to the south but failed because the artillery and the trains could not negotiate the combination of bad roads and deep snow. A second attempt also failed. Thereupon it was decided that the infantry should advance to the south accompanied only by a few guns drawn by double teams.

To leave the bulk of the artillery behind may strike the reader as dangerous but I believe the decision to do so was correct. The Germans were pursuing and almost anything can be dared where opposed to a beaten opponent. Everything had to be sacrificed to speed if the Russians were to be overtaken. In this situation legs were the important thing, not cannon.

Long difficult days of marching followed. These days were particularly trying because the regiment was without field kitchens for a week. Finally it came into the area just north of Suwalki. There, after a short skirmish with Russian flank guards, the regiment again bent to the east, frequently marching at night. Ultimately, it arrived in the rear of the Russians who were then inclosed on all sides by German troops.

For a few days the troops of this regiment found themselves in a peculiar situation. The inclosed Russians made

[44]

continuous effort to break out of the forest of Augustowo toward the east while from the fortress of Grodno attacks were made in a westerly direction to assist them. The Germans stood in the middle, and therefore had to fight toward two fronts. While this regiment, reinforced by artillery, was fighting with its front to the east facing Grodno, the rest of the division stood with its front to the west facing the forest. For days it heard the thunder of hostile cannon at its back while facing an enemy to its front. This was not a pleasant situation. During this period, however, the field kitchens returned, an event that added immensely to the morale of the regiment.

One bright afternoon, the Russians attacked from the direction of Grodno. The ground was covered with snow and, at a distance of 1100 yards, the Germans coud pick up every individual. It was impossible for the Russians to find cover anywhere. Every movement they made was observed. Their attack was repulsed with heavy losses. On the following morning, they again attacked but this time in a thick snow storm. This time they appeared right in front of the Germans before they were seen. The element of surprise being with the Russians, they were beaten off only after serious fighting.

In the meantime, the Russians in the forest had been forced to surrender. About 90,000 prisoners and 200 guns were captured. The Russian Tenth Army was destroyed.

In analyzing the reaction of these young soldiers under the varying conditions of this campaign, four important lessons stand out.

1. First, they had a commander whose very name inspired confidence. We shall seldom have, at the beginning of a war, a proven commander, but we should have one who can inspire the confidence of the troops. It must be the aim of every commander to gain such confidence as quickly as possible. If he does, he can demand anything from his men.

2. The German troops were young and had only undergone a short period of training. They had entered the army in October, 1915 and therefore had but three months of training behind them. However, they were intermingled with men who had already had some war experience, and

who at least knew those first impressions that war brings. These veterans regarded themselves as instructors to their young comrades; they felt a certain responsibility for them. Because of this feeling the value of the old soldiers was markedly increased, while the inexperienced men developed rapidly under their instruction. Although only one-fourth of the men were experienced, their influence was sufficient to give the entire organization a veteran character. I watched this same influence at work later in the war and it always had the same result. For example, in September, 1915, at the end of our great Russian offensive, the companies of our battalion were reduced to only twenty men. Each company received from forty to fifty replacements, and a few days later twenty more. After several days of marching, we again went into action. It was almost impossible to differentiate between the old men and the new. The spirit of the company seems to instill itself automatically in the recruit. They have confidence in their more experienced comrades and learn very quickly from their stories and from their actions.

What do we learn from these facts for the future? According to my view there is one tremendously important thing. At the beginning of war new troops will be recruited and trained in all countries and naturally will enter into combat later than the active troops—frequently months later. If we give these inexperienced troops a backbone of experienced soldiers and experienced commanders their efficiency will be tremendously increased and they will be spared heavy losses.

3. It will be recalled that on that foggy morning when the battalion was suddenly halted its commander issued a short but adequate order, although there was no information of the enemy. This battalion commander had learned from his recent experience in the war. He knew that lack of knowledge of the enemy is a normal thing. He had learned that absence of information does not justify withholding an order when an order is needed. The result was that the troops deployed and were ready for combat when they met the enemy.

4. Finally, you saw that the first German attack and the Russian attack in the snowstorm were effective because of the surprise factor. In the former, success was gained without

heavy losses and while the Russian attack failed, it was beaten off only with difficulty.

The importance of surprise in war cannot be overestimated. As it becomes increasingly difficult to obtain so does it become increasingly effective when it is obtained. No effort should be spared to make the decisive element of surprise work for us in war.

CHAPTER V

A Young Company Commander in a Rapidly Changing Situation

(This article appeared in the July-August issue of the Infantry Journal)

It was about the 20th of September, 1915. For some days a German battalion had been marching over bad roads, deeper and deeper into Russia. During the middle days of September this battalion had been heavily engaged against the Russians in the vicinity of Vilna. Afterwards it resumed the march to the east. For several days no organized enemy had been seen. The marches became increasingly tiring as the battalion had now come into a land of swamps, forests, and sand.

The men did not know whether they were marching as part of the division or as part of the regiment nor did they care. In peace, soldiers are taught that even the last private should be acquainted with the situation; in war, no one knows anything about it and generally has no wish to know anything about it. This lack of interest seems to be one of the main reasons for the fact that the rank and file know so little of the situation. The soldier is satisfied if things go well with him. He has enough to do to look after himself; he must march, sleep, eat, and when he meets the enemy he must fight. Whether the entire division is attacking or only the battalion is generally a matter of complete indifference to him. Such was the case here. The soldiers marched, slept, and ate. Where they were marching, or what they should do, was of no consequence to them. Their higher commanders could make all the strategical plans they wanted—the soldiers marched. It is true that they frequently asked each other what their mission could be. Then someone would reply "We are marching to Moscow". A laugh would follow and everyone would be satisfied.

On one of these autumn days, at about ten o'clock in the morning, the battalion suddenly bent from its previous direction and turned south. The route of march followed narrow forest paths, and more often than not ran across country. Suddenly a couple of messengers galloped up, spoke briefly with the battalion commander and then dashed off. Something must be wrong. The battalion commander had ridden forward.

Suddenly he returned, saying: "The Russians have attacked our cavalry with strong forces and pressed them back. We are to go to the cavalry's assistance. About two kilometers in front of us is a river. We will hold the river line."

The river was soon reached. It was about 40 yards wide and very deep. On the far bank was a village. On our side the woods and fields lay in an irregular pattern. Not a shot could be heard. Not a Russian could be seen. Not a German cavalryman was visible.

The battalion commander had the only available map. He studied it for a few moments and then turned to his company commanders: "Over there on the right about 500 meters away is a farm; a battalion of the X Regiment will be there. We will defend from this edge of the wood to the left. The 9th, 10th, and 11th Companies, each with a sector 300 meters wide, will hold the front line in the order named from right to left; the 12th Company will be held in reserve behind the center of the battalion. Cavalry is operating on our left. We have no artillery for the time being. Send patrols across the river. I will get in touch with the cavalry."

Let us go with the 9th Company. The young company commander led his company to bivouac in the wood located about 200 meters from the river and then in company with a few men moved forward to the river in order to look the situation over and see the real lay of the ground. Before he left his company he sent a patrol to the right to gain contact with the battalion at the farm.

All was quiet at the river. The situation did not seem as bad as the battalion commander had pictured it, although the terrain was very unfavorable. The ground was so thickly wooded that visibility in any direction was limited to about one hundred meters.

The company was about eighty to ninety men strong. The company commander decided to use two platoons along the river in the front line and to hold out the third as a reserve. He issued orders accordingly.

His platoon commanders, although only corporals, were men on whom he could rely. They returned to their platoons while the company commander searched along the river for a boat in which he could reach the far bank. Although he wanted to see what it looked like on that side, it must be admitted that his principal reason was the hope that he might find something to eat in the village across the river. Finally he found a boat. As he did so he looked back and saw his platoons coming forward from the wood. All was going smoothly.

Suddenly, over on the right, a rifle was fired. He thought to himself that someone in the battalion at the farm was killing a pig for his field kitchen. Now he heard another shot.

Again he thought to himself: "Ah ha! a bad shot. He certainly ought to have been able to kill the pig with his first shot." Firing increased, two, four, seven shots.

"Can there be a fight over there?"

The firing now became heavier. Quicker than can be told the following events, thoughts and decisions took place.

The company commander's first thought was: "The neighboring battalion has located and driven back a Russian patrol. The firing, however, seems to be too strong for a mere patrol fight."

In the meantime, however, he climbed out of the boat.

"My patrol will certainly bring me information," he thought.

Suddenly a few rifle bullets whistled over his head, coming from the right rear. By their sound he knew that these bullets came from Russian rifles. The situation now became clear to him. The neighboring battalion was not at the farm, otherwise, the bullets could not have come from the right rear. The Russians had crossed the river and were at the farm!

What should he do? The battalion commander was not there. He had to make his own decision. He had to act and

[50]

at once! Naturally these thoughts did not require the time it takes to relate them.

His train of thought was perhaps as follows: "Mission, defense of the river. The situation is changed. The Russians are across the river. Therefore, my decision is to attack quickly and with as much strength as possible."

He ran back with his runners to the reserve platoon. On the way he gave this order to a runner, a man in whom he had complete confidence:

"The left platoon will immediately retire to the wood and will then follow me in an attack on the farm. The right platoon will defend the entire company sector. Deliver this message to the platoon leaders and then report my decision to the battalion."

He was soon with his reserve platoon which had taken position with its front to the farm. However, the men could see nothing. Firing was still going on. Without halting a moment, the young commander yelled:

"The whole platoon will attack in double time in the direction of the farm."

The whole outfit plunged into the wood at the double. It was necessary to attack the Russians quickly before they could get across the river in large numbers. A messenger came running up from the patrol breathing hard:

"The Russians are across the river near the farm. The patrol is lying down along a little road which leads north from the farm. The Russians are trying to get around us."

A new situation had now arisen. Since the first decision scarcely ten minutes had passed. The company commander's thoughts ran something like this:

"Has the situation changed? Do I now have to make a new decision? Is it possible to continue the attack? Are the Russians already across the river in force?"

The company commander had only thirty men with him. He decided to attack. It cannot be said whether this was right or not but such was his decision.

The advance continued. Soon the thinning undergrowth indicated that the edge of the woods was near. Just before the assaulting troops was another wood and a road leading to the north. Russians were on this road.

"Lie down, range 400 meters, fire."

That was the only order given. Fire broke loose. For a moment there was no reply from the enemy, but soon a hail of bullet came down on the Germans. It was evident that a mass of Russians were concealed over there. At this moment a runner came from the 2nd Platoon:

"The 2nd Platoon is 200 meters in rear of us."

Again the company commander had to make a decision. Should he continue the attack? No time was available for long thought. He called out:

"I am attacking with the 2nd Platoon on the right. This platoon will keep up its fire and then join the attack."

He ran back to the 2nd Platoon and led it forward through the wood toward the right. As they ran he issued his order:

"There are Russians near the farm. We are attacking."

In a few moments the 2nd Platoon had also reached the edge of the wood. As they emerged from the wood they received heavy fire from their right flank, which forced them to take cover.

The Russians were already much farther across the river than the company commander had believed possible. The situation had again changed. What should he do? Would he have to make a new decision? At this moment a runner came from the battalion:

"The Russians have broken through our cavalry. The battalion commander is wounded."

The runner could not say a word more for he sank to the ground dead with a bullet through his head. Again the situation had changed. What should be done? Would a new decision have to be made?

Naturally the situation was not as clear as it appears in the telling. The company commander did not have a map. He stood in the midst of combat. One thing, however, seemed clear. To carry on the attack with only fifty or sixty men would surely lead to failure. But what should he do? There were only two possibilities—hold where he was or retire.

Defense would be advantageous only in the event that fresh German troops were available who could drive the Russians back across the river. The 12th Company was still in reserve, but no one involved in this fight had seen or heard anything about it. The company commander therefore came to the conclusion that it had probably been used on the left flank of the battalion where the cavalry had retired. There still remained one other body of troops that might be looked to for help—the battalion that was supposed to be at the farm. It was not there. When would it arrive? Would it come at all? No one knew. The Germans could hold their own if they received support immediately. If no support was coming it was high time to retire. To remain where they were, fighting without prospect of immediate support, was equivalent to destruction. Under these circumstances he decided to retire.

It is immaterial here whether this course was right or wrong. The essential point is the fact that this decision had to be felt rather than arrived at through logical thinking. If the battalion which had been awaited so long had arrived in the next few minutes, this decision would have been a serious mistake. However, the matter of prime importance was no longer a question of holding the river, but of holding the enemy. If he could not be stopped at the river, then he had to be stopped at some other place.

The retirement was made slowly. In the meantime the Russian artillery opened up. By afternoon the company had regained contact with its battalion which had in time been joined by the cavalry. Later another battalion arrived and a few batteries of light artillery put in their appearance. The crisis was over. A defensive position was organized on a little ridge in the midst of swamp and forest and during the next few days every attack the Russians launched was repulsed with heavy loss.

During this fighting an incident occurred that is interesting because it shows how often war demands unique methods. The line was very thin in this defensive position and the men widely scattered. This battalion's reserve consisted of only thirty men who were called on every time the Russians attacked. Companies lay in the swamp without trenches and without wire. The terrain was such that they could see only a short distance to the front. Everywhere there were little bushes and patches of woods that obscured the general view. There were so few men that strong outguards could not be sent out. Therefore, it was often most difficult to determine when the Russians were going to attack. They were generally seen for the first time only after they were close to the position.

What could be done to remedy this state of affairs? A corporal of another company hit on a simple and excellent solution. About 500 meters behind the lines there was a village which boasted a large number of cows. The corporal went to this village and drove all the cows in front of the position where they could quietly graze. As soon as the Russians put in their appearance, the cows became uneasy and moved toward the German lines, thereby giving warning that a new Russian attack was imminent. These cows were the most effective outpost that could have been found.

Let us summarize the more important lessons to be learned from this action.

We saw in this fight a complete absence of information of the enemy, and a lack of knowledge of the situation. Although the German cavalry had been in contact with the enemy and was probably in a position to give information, nevertheless no information materialized. Suddenly the Russians were across the river. What their strength was is not known to this day. Perhaps, in future wars, the approach of the enemy will be discovered by aviation. However, we should not expect too much from air reconnaissance. Both sides will have aviation at their disposal. Before our aviators can make a good reconnaissance, the hostile air force must be thoroughly whipped. If we can gain and keep supremacy of the air, we may get considerable information, but on the other hand if the enemy is victorious in the air, we can expect little or nothing from this source. Furthermore, in open warfare prior to combat aviation will obtain information principally of the large units of the enemy—marching columns, direction of march, etc. This information will reach the higher staffs in the rear. But what we in the front line want to know about the enemy—the location of his machine guns and his centers of resistance—our fliers can tell us only rarely, and then as a rule too late to do us any good.

Moreover, we infantrymen will constantly endeavor to conceal ourselves from aviators by marching at night and by splitting our forces into little groups. If such are to be our own tactics, surely we must expect the enemy to conceal his movements by similar methods.

Thus it appears that in the war of the future we will again be required to make decisions without satisfactory knowledge of the enemy. It is therefore important to practice this in peace for in war we will do well only that which we have learned to do in peace.

This example shows clearly that difficult situations can be solved only by simple decisions and simple orders. The more difficult the situation the less time there will be to issue a long order, and the less time your men will have to understand it. Moreover, the men will be high-strung and tense. Only the simplest order can be executed under such conditions.

The situation pictured in the foregoing narrative developed slowly at first, then suddenly difficulties come thick and fast. As we look back on the war, we will find this true in almost all situations. From this we should draw several conclusions with respect to our issuance of orders. The first order for the fight can generally be given without hurry. It should contain everything that is necessary. However, once a fight has started, we cannot expect to be able to issue long-winded orders, either written or oral. Whatever order we are able to issue must be short and must be clear. If we hope to do this in war we must practice it constantly in peace.

And now the last point, and the most difficult of all. Our map problems generally close with a statement that it is now such an such an hour and call for a decision. We know, therefore, that the situation is such and such, that we have all the information we are going to get and that we must make a decision. The foregoing action clearly indicates that one of the most difficult things we have to do in war *is to recognize the moment for making a decision.* The information comes in by degrees. We never know but that the next minute will bring us further information that is fresh and vital. Shall we make a decision now or shall we wait a little longer? It is usually more difficult to determine the moment for making a decision than it is to formulate the decision itself.

Here is one method by which an organization may be

trained in this matter. Go out on the ground with your platoon commanders, your section leaders, your squad leaders. Tell them:

"You are marching with your platoons on this road."

After marching a while, give the leaders a bit of information or an impression. March quietly on and then give them another bit of information. Build up your problem in this manner. Never ask them for a decision. Tell them that they must decide not only what to do but when to do it. You will then see how terribly difficult it is to determine the time to make a decision.

Let it be emphasized again that we must learn all these things in peace. We must not only know how to do them in theory, we must know how to do them in fact.

Let us close with a thought from General von Seeckt: "The knowledge of the trade is essential, but it is the work of an apprentice. The task of the journeyman is to utilize what he has learned. But the master alone knows how to handle all things in every case."

CHAPTER VI

Contrasting Service with Trained and Untrained Troops

In previous chapters it has been shown how young troops react to all new impressions. It has also been shown that the value of the new soldier depends on his leaders to a far greater extent than that of the seasoned veteran. Furthermore the fact has been stressed that the confidence of the soldier in the experience, the cleverness, and the ability of his commanders, helped him to surmount the difficulties that confronted him in his first operation against the enemy.

In all the actions previously discussed, our troops were on the offensive and we were successful. Attacking troops are imbued at the very start with a certain sense of superiority and confidence in their own power. This is a tremendous assistance to them in overcoming the demoralizing impressions of combat.

I have had no experience with young troops on the defensive or during a retirement. It is my belief, however, that a poorly conducted retreat, or a defensive action which fails, will reduce the value of young troops for a long time. On the other hand, veteran troops who have undergone the vicissitudes of many campaigns do not succumb so easily in adversity.

Generally, the soldier at the front thinks only in terms of the present; he lives from day to day. If a combat action is successful, he is pleased. He is not concerned with the great strategic results of battle; his thought revolves only about himself and his immediate surroundings.

In preceeding articles I have commented extensively on the lack of enemy information so characteristic in war. I now propose to develop another subject whose lessons are equally valuable: namely, a consideration of difficult and unexpected situations.

Looking at the war in retrospect it seems to those of us

who were at the front that most situations were difficult and unexpected. The rules with which we solved them were generally simple, and almost invariably at odds with the standard solutions we had learned in peace.

In early October, 1915, the Germans were deep in Russia. For several months there had been much fighting and marching. Then came a week spent in heavily wooded and swampy terrain. Our soldiers, veterans of many battles were now hardened to war and its surprising situations. They knew their leaders and their leaders knew them.

After several days of heavy fighting, an unexpected order arrived: "Retire to the north". No one knew anything further.

During the night contact was broken with the enemy. A thirty-hour march to the north followed. Let us accompany one of these German battalions—the rear guard of a division out of contact with the enemy.

Late the next morning an officer arrived from the division staff and delivered the following oral order to the battalion commander:

"The division will rest for a day along the route of march; its rear elements will be at such and such a village. Your battalion, as rear guard, will provide security for the division in the vicinity of the village of X. The march will not be continued until to-morrow night. A troop of cavalry which is still behind us is attached to you. Two companies from the 1st Battalion will provide security on the adjoining road to the west. A written order follows."

The battalion commander said: "All right, my dear fellow, but couldn't you tell me where the village of X is? I haven't a map." The staff officer produced his own map, indicated the village and promised to send a map.

After an hour's march the battalion reached the designated village.

This battalion consisted of four rifle companies and one machine-gun company, but each company was only thirty men strong, and the machine-gun company had only two guns left. These two machine guns were carried on a peasant wagon which was drawn by two little Russian horses. The battalion commander was the only officer in his proud command. His staff consisted of a sergeant, two mounted messengers, and his striker. He had appointed the sergeant as his adjutant, because he wanted an adjutant above all else. Of the two mounted messengers one had been sent to the division staff to obtain orders, the remaining one accompanied the battalion commander. Three of the companies were commanded by sergeants; one by a corporal. But the battalion still had one great source of strength—four field kitchens, which were used daily just as if every company were still 120 men strong.

The men were veterans who couldn't be gotten out of their habits of peace and quiet. They had marched for days and they were hungry and tired—terribly tired. Nothing was known of the enemy. Put yourself in this young battalion commander's place. What would your decision be? This was his. He assembled his company commanders and issued the following oral order:

"Our division is bivouacking over there behind those woods, at least until to-night. We will protect them. If the Russians come we will defend ourselves on this ridge. 9th Company in the middle, 10th Company on the right, 11th Company on the left, machine-gun company on the right flank. The 12th Company will take position behind this hill in reserve. I will be with the reserve. I am going on a ride to reconnoiter the terrain. The battalion will go into bivouac for the time being. The 9th Company will go into this house, the 10th Company into that house, and so on. Get some sleep and see if you can find something to eat. The 9th Company will send a patrol about two kilometers back into the woods in the direction from which we have come. If the Russians come, this patrol will fire. I will be back in about half an hour."

Then he started on his reconnaissance. I am glad that this was no marked problem for this young battalion commander would certainly have received a "U." But think a moment. His men were veterans who knew that ordinarily a situation does not develop very rapidly. They were tired and they were hungry. The battalion commander, of course, had no idea where the Russians were. He didn't have a scrap of information about them. He only knew, that over there toward the west, 250 miles away, was

Germany, and that his division was retiring in a northerly direction. However, he knew his men, and knew what he could ask from them. Not knowing what the next few hours would bring forth, he was not worried about the future. His men had come through many difficult situations and he was confident that they would come through this one, particularly in view of the fact that no difficulties were in evidence at the moment.

After half an hour he returned. The entire village lay still and dead—nothing moved. The striker was asleep in front of a house. The companies were snoring inside the house. One double sentry post stood well concealed in the south edge of the village and observed the forest. The striker had prepared something to eat. It was about two o'clock in the afternoon.

Think of this strange situation! A division is resting during its retreat and has charged a battalion, which consists of only 130 men, to secure its safety. This battalion sends out a patrol, establishes one sentry post at the edge of the village and goes to sleep. Compare this situation with the one in the early days of the war, when 80 excited men guarded 160 other men during their rest.

It was probably about three o'clock when the battalion commander was awakened. "The Russians are coming, our patrol is firing." The companies were quickly awakened and in a few minutes they were in their defensive position on the ridge. Suddenly a troop of German cavalry dashed out of the wood. Several shrapnel shells burst above them. The patrol was still firing but nothing could be seen of the Russians. The German cavalry galloped over the infantry position and disappeared to the rear. The battalion waited for something to happen. Firing was still going on in the woods to their front.

The battalion commander mounted his horse and rode to the right in order to establish liaison with the companies of the 1st Battalion. On arriving there he saw no movement. He rode farther. There wasn't a single German anywhere in the vicinity. Then he suddenly saw several Russian cavalrymen in the edge of the wood.

It was now about 4 o'clock. At six the sun would go down. What would he do? In this new situation there was no

[60]

time to make a long estimate. He had to act. He turned to his mounted messenger: "You remain here and fire at everything that appears at the edge of the wood. I will ride back to the battalion and send infantry." Then he galloped off.

On the way back to his battalion he thought the situation over: "The Russians may attack us any moment. The companies of the 1st Battalion have disappeared. A little while ago they were over there. Surely they have withdrawn under orders. No order to retire has reached me. A decision must be made." He decided to withdraw as quickly as possible. But where? He had neither map nor compass. He only knew where north lay. "Well," he thought, "let's retire first. After that we must have some luck." With these thoughts he came back to the battalion and gave an order: "Companies immediately assemble behind the ridge. The battalion reserve and the machine guns will protect the withdrawal."

Just as the runners were about to leave wtih this order, the patrol, which had been in front, came out of the wood at a dead run. Soon thereafter the first Russians appeared and opened fire. The situation had changed again. A withdrawal could not be executed now until the Russian attack was beaten off. Promptly the battalion commander sent half of his reserve to protect his right flank. The Russian advance was very slow. Their small arms fire gradually became stronger and a few shells fell, but little damage was done. The Germans shot rapidly and only when they had a good target. Night began to fall. Twice the battalion commander rode over to his flank security group. They reported that the only evidences of hostile activity to their front were a few Russian cavalry who showed themselves from time to time.

Finally, after darkness had completely closed in, the battalion withdrew. Before the withdrawal began, the battalion commander told his men that he did not know where the division was and that their chief task lay in marching until they found it again. The battalion moved off in the night. There was no moon but the stars were out and indicated in which direction north lay. Now and again they came to a road fork. The battalion commander, not knowing whether they should march to the right or to the left, let his horse decide. If he went left, they marched left, if he went right, they marched right. Through forest, through sand, through swamp they went. Now and then they passed through lifeless villages, dead and black in the gloomy forest and the dark night.

The battalion commander was uneasy. Occasionally he rode forward. Again he hung back at the rear of his column. Sometimes he rode beside his silent men, who moved rapidly along the dark road like shadows. Often the only sound heard was the low moaning of four wounded men who were being carried along on the peasant wagon. Now and again a soldier cursed as he fell in the darkness. At intervals a short halt was called. "Were we on the right road? Would we meet our division? What should we do if we were unable to find our division?" Such were the leader's thoughts. . . . Suddenly the voice of one of the soldiers came out of the darkness, "What shall we do if we don't find our division?" The commander answered: "Boys, then we will make our own war. We shall march westward until we meet Germans again. No one will take us prisoners." A little laughter followed, then complete silence. Commander and troops were one in their thought.

At three o'clock in the morning, after nine hours of marching, the battalion crossed a long bridge over a lake. A large village lay on the other side. Here was a fine place for a halt and the commander gave the word. Within five minutes the entire battalion was billeted in a single large house. Five minutes more and all were sound asleep. A machine gun and crew were placed on the bridge as a guard. The battalion commander directed that the march would be resumed at six o'clock. Then he fell asleep.

Suddenly he awakened. The sun was high. It was 8:30 AM. He ran out of the house and down to the bridge. There lay the entire crew of the machine gun sleeping. At that moment he saw, on the farther side of the bridge, 300 to 400 yards away, a troop of Russian cavalry. He woke the gun crew roughly and, just as the first hostile cavalrymen started across the bridge, the gun opened fire. Several Russians fell. The survivors fled.

This firing was the best alarm clock possible for the

battalion. Within a few minutes the troops were again on the march northward. About evening they found their division, which had halted. Here the battalion commander learned for the first time where he was. He was east of the Beresina, that river on which Napoleon's proud army had met disaster in 1812. He also learned that the soldier who was to have brought him the order for withdrawal had lost his way and his horse in the swamp and had finally returned without delivering his message.

This narrative has not been told merely because it makes interesting reading, but primarily because it brings out several important points.

In our peace-time map problems, war games and field exercises, we have simple situations. There is no uncertainty, nothing goes wrong, units are always complete. Every company has its appropriate number of officers. Every battalion has its commander, its executive, its S-1, S-2, S-3, S-4. Long written orders are published and in an unbelievably short time, reach the individual to whom they are addressed, who promptly carries them out. Every man has his map and compass. He knows that the attack will be pushed forward in the exact direction of 179 1-2 degrees magnetic azimuth.

In war it is quite otherwise. There is no situation that our imagination can conjure up which even remotely approaches reality. In peace we have only grammar school tactics. But let us never forget that war is far more advanced than a high school. Therefore, if you would train for the realities of war, take your men into unknown terrain, at night, without maps and give them difficult situations. In doing so use all the imagination you have. Let the commanders themselves make their decisions. Teach your men that war brings such surprises and that often they will find themselves in apparently impossible situations.

I have spoken earlier of surprises of a psychological nature; in this article I have talked of surprises of a tactical nature. Every soldier should know that war is kaleidoscopic, replete with constantly changing, unexpected, confusing situations. Its problems cannot be solved by mathematical formulae or set rules.

All armies of the world learn, in peace time, how to

write long, beautifully constructed orders. I believe that it is correct to learn to think of everything and to forget nothing, but we must never lose sight of the fact that, in a war of movement, our orders must be brief and simple.

There is a tendency in peace time to conduct training by use of stereotyped situations which are solved by stereotyped solutions. In war, however, we cannot say, "This situation is so and so and according to the rules which I have learned, I must attack or defend." The situations that confront one in war are generally obscure, highly complicated and never conform to type. They must be met by an alert mind, untrammeled by set forms and fixed ideas.

In our peace-time tactical training we should use difficult, highly imaginative situations and require clear, concise, and simple orders. The more difficult the situation, the more simple the order must be. Above all let us kill everything stereotyped; otherwise it will kill us.

CHAPTER VII

The Borderland Between Open and Trench Warfare

So far we have considered only open warfare with its attendant uncertainties and surprises. Both opponents have been in movement. Trenches and prepared positions have been non-existent. Detailed reconnaissance has been impossible; the location of hostile machine-gun nests, artillery, and centers of resistance have been unknown.

Now, however, we arrive at a new phase of warfare; the struggle of the defeated enemy to bring about stabilization, to pass from warfare of movement to "war in a slot." He bends every effort to accomplish his end; he brings up fresh forces, he organizes new defenses, and deeper and deeper, he burrows into the earth. To all armies have come those sad days that we experienced when we struck again and again, striving to resist the impending threat of trench warfare, never knowing whether or not we could once more bring fluidity to a slowly freezing front.

Thus, the final period of that phase of warfare which we term open, is characterized by fierce struggles for individual trenches and terrain features. The World War has taught us that almost every period of open warfare is temporarily broken by periods of threatened stabilization. This threat arises when a beaten enemy seeks to organize and hold new positions while the aggressor strives to push his advance still farther. If the attacker be victorious, warfare of movement is resumed; if he fail, the battle stagnates in the trenches.

To the infantryman, a consideration of such battles is of vital interest for it is his lot to bear their heaviest burden. In fact, the issues in such engagements are decided by the infantry alone. Let us consider this briefly.

Open warfare consumes an army's strength to an unbelievable extent. Battle casualties pile up. Sickness increases in direct proportion to the duration of the campaign. The losses in animals become even greater than those in men. Supply lines become longer and more difficult. Heavy artillery lags behind. Losses of guns, trench mortars and machine weapons become increasingly serious as the problem of replacement grows. Day by day the difficulty of obtaining ammunition for the still existing artillery becomes more acute. Gradually, we see the military machine, as a machine, grow less and less important while the human being, the infantryman with his rifle, becomes more decisive. Though his numbers decrease and his physical strength diminishes daily, it is the infantryman who must increasingly bear the burden of battle.

Before considering several engagements of this type, that occurred during the World War, let us review some of that same war's lessons, albeit, no human being can say whether or not they will apply to the war of the future. Indeed, there is tragedy in the fact that the soldier must learn from examples of the past and only rarely from the present. In fact, there is a certain danger in the study of military history if we seek to obtain from it more than the eternal verities of leadership, morale, psychological effects, and the difficulty and confusion which battle entails. We cannot visualize war of the future merely by studying wars of the past. If this were so, the best professor of history would necessarily be the best commander. Certainly the next war will place entirely new problems before us—problems that have not even been imagined. Our descendants may never have to face the difficulties that confronted us in the last war and that still confront us today, but we can be certain that in place of these they will have other difficulties to overcome.

In considering the events we shall speak of, it should be borne in mind that the lessons they contain are based on the World War, which now lies about twenty years behind us. Let us not fall into the mistake of definitely expecting the same lessons to be derived from the next war. Perhaps they will be the same, perhaps not.

We now turn to a consideration of certain engagements in that dim borderland between open and trench warfare, in those days when the offensive power of the temporarily victorious side began to weaken while their opponent's defense grew stronger.

The Russians had been steadily beaten in 1915 and the Germans had been pursuing them for months. Daily, in rain or shine, there were long, tiring marches across the wide Russian plains, through broken hill land, through enormous forests, and mostly on bad roads. Almost daily there was fighting—sometimes large engagements, sometimes skirmishes. All of these brought losses—losses in men, in horses, in material. Companies and battalions became smaller and smaller, but those men who remained were war-experienced veterans with iron bodies and iron nerves, upon whom their leaders could implicitly depend.

During the last few days the Russians had been offering opposition more and more frequently and with increasing energy; their artillery had become more active and they had even undertaken small counterattacks.

The battalion we are accompanying was marching in the main body of its unit. Suddenly a fusilade broke out forward. The battalion halted. There was a brief delay— a few moments, then it marched a few kilometers farther. Again there was firing, again a halt. "Ah, to-day we are lucky to be in the main body," thought the soldiers, "We are agian faced by a small Russian rear guard, but our advance guard will quickly bring everything into order."

Although autumn was already on us, the weather was still sunny and warm. "Halt" was no sooner given than the soldiers would throw themselves on the ground and sleep. What mattered the firing up forward? Here was an opportunity to sleep, so they slept.

Thus the battalion moved forward by bounds. About ten o'clock in the morning the firing to the front became heavier. The battalion bent abruptly from the route of march and, utilizing ravines and patches of woods, continued to push on. Soon we heard the deep reverberation of artillery and the occasional rattle of machine guns. At last we halted in a small wood while the company commander rode forward to join the battalion commander.

Order for the attack: "The Russians are in a position about four kilometers in front of us." The battalion commander pointed to his map. "They are well dug in on those hills behind that broad swamp. Before their position is a wide, barbed wire entanglement. Our division attacks at 2:00 PM. The zone of attack of this battalion is—so and so. 9th, 10th and 11th Companies in front line. Boundaries between companies—this and that. The 12th Company, behind the center, will be my reserve." Such, in effect, was the attack order.

The battalion took up the approach march. At first the companies moved forward rapidly, but before long artillery fire located them and they were forced to deploy. Still nothing could be seen of the enemy, though his bullets whistled over the hills and crackled in the trees. From one of the nearby hills, the young company commander of the 9th Company attempted to locate the Russian position. For a long time he was unsuccessful. Suddenly a soldier in his company cried out; a few seconds passed and another one screamed. They had been struck by infantry bullets but the hostile position remained undisclosed.

At last the company commander discovered it. Almost perfectly concealed, it lay in the hills well to his front. It must have been one of those famous Russian positions laid out long ago, for there were no evidences of fresh digging or deep trenches covered with earth and boughs such as the Russians built in early 1915. Instead of this, he saw a barbed wire entanglement ten to fifteen meters broad and in front of this a grassy swamp about two hundred yards wide. A low ridge ran along the German side of the swamp. How could this strong position be best approached?

German artillery was only firing occasionally and even these few shells were flying far over into the Russian rear areas. Perhaps they were seeking the hostile artillery. The Russian position itself gave no indication of life. The company commander rose to cross the hill on which he had been lying, but scarcely had he gained his feet when the Russian trenches became alive. Rifles cracked; machine guns rattled!

What could be done now? To launch the attack from the hill on which he stood would serve no purpose. It was much too far away. It was essential to get close to the

Russians. The first problem was to gain the low ridge that ran along the German side of the swamp. From his position on the crest, the company commander located a deep ravine in the side of the hill. He decided to advance through this ravine by squads.

The advance proved slow and tedious. The Russians fired at every German who showed himself. Nevertheless, the company got forward successfully, sustaining only eight casualties. It was 1:30 PM, however, before the edge of the swamp was reached. Only thirty minutes remained before the attack. The company was quickly organized in its assault formation.

But where were the neighboring companies? Time was no longer available in which to search for them but patrols were immediately sent to the right and left to endeavor to gain contact with them. Meanwhile the Russians continued firing at every individual who showed himself. Our artillery was firing only occasionally.

Just before two o'clock the telephone arrived. Over this, the company commander reported to the battalion commander and questioned him about artillery support. "I don't know anything about it," he replied.

The hour of attack was approaching. Before them, the Germans saw a few Russian trenches, but they did not know where the hostile machine guns were located. They saw the wire entanglement but they did not know how they could force their way through it. They saw the broad meadow marsh but they did not know how they could cross it. They had no machine guns and they did not know whether any artillery was available to support them. However, in the past months these troops had made so many attacks, in every one of which they had been victorious, that they now believed unconditionally in their superiority over the Russians. They believed to a man that their enemy would offer slight opposition for a short time and then retire. They were as sure as fate that they would be victorious again today.

At two o'clock the attack jumped off. Scarcely had the Germans shown themselves on the edge of the marsh when hell broke loose from the Russian lines. Their entire trench system swarmed with life. Infantry, artillery and machine-gun fire swept through the attacker's ranks like a tornado. Heavy casualties occurred instantaneously and we were immediately forced to take cover. Heavy casualties were something the Germans could no longer afford. The combatant strength of the individual companies was already so reduced that the 9th Company had been unable to keep out a reserve. The attack was repulsed almost before it started. Losses were severe and nothing had been gained. It was impossible to close with the enemy.

The fight had now begun along the entire front. To the right and to the left, German soldiers advanced but not one of them succeeded in crossing the swamp. So passed the afternoon. It became dark.

The company commander of the 9th Company planned to come to grips with the Russians under cover of darkness. With twenty men he crawled into the swamp. He actually succeeded in reaching the Russian wire but scarcely had he started to cut through it when a storm of fire again broke loose. Directly in front of the small group of Germans and not more than fifty meters away a machine gun was firing. God be praised! It was so sited that it was unable to fire at such a close range. A hail of bullets poured from the trenches. The Germans flattened themselves on the ground but even so the bullets cracked and whined appallingly close to their heads, striking the marshy ground in rear with vicious little plopping noises.

It was impossible to get forward and equally impossible to withdraw. Any man who raised his head, even an inch or two, was as good as dead. Unfortunately, several men of the little group did raise their heads. Desperately the others tried to dig in. With hand and knee and even mouth, they scooped out the earth until they had secured some degree of cover.

The Russian fire now drew a reply in kind from the German lines on the other side of the marsh. Thus the little group was exposed to fire from their own comrades as well as from the enemy.

Almost at the very muzzle of the Russian machine gun, they passed the night—quiet, motionless, waiting for the death that daylight would certainly bring. Any one who has been exposed to dire peril for a long time, without being able to oppose it, can understand what a terrible nervous test such a night imposed on this handful of soldiers.

The stars had dimmed and the east was beginning to grey before the Russian fire subsided. The German fire had died out some time before. Individually, the few exhausted survivors crawled back to their company on the other side of the marsh. But, of the twenty who had gone forward only four returned. The others lay cold and stiff before the Russian wire. The attack of the battalion had failed.

The new day was utilized in digging in and in establishing liaison to the flanks. Even a number of dugouts were constructed. Once more it seemed that trench warfare was about to begin. But two days later the thunder of a great battle was heard far to the right. It lasted during the entire day. The next morning German patrols found the Russian position to their front deserted. Again the Germans had successfully brought back open warfare; and again began the long marches over hills, through valleys, through seemingly endless forests.

This little segment of a battle is typical of fighting in the borderland that exists between open warfare and trench warfare. The attacker sees before him an enemy who is preparing to offer fresh opposition. He doesn't know whether it is to be another fight against a rear guard or whether the enemy has finally halted. If it be a rear guard, the enemy is only trying to gain time. In this event the attacker must strike quickly to deny him that desired time. On the other hand, if the enemy is no longer willing to retire, it is not probable that he can be driven back by a quick attack.

The first lesson to be drawn from this phase of combat is its uncertainty. We know that the enemy is defending a definite place, but we do not know his strength and above all, we do not know his intention. In the future, aviation may be able to clarify this but it is certain that the more efficient air reconnaissance becomes, the more baffling will be the measures of defense. As the aviators' eyes reach farther and see more clearly, troops will retaliate by concealing themselves with greater skill. *The best reconnaissance will always be the attack.* Often it will be the only device which can reveal the strength, dispositions and intentions of the enemy. Visual reconnaissance and patrols, of themselves, have never been able to obtain this information in the past and probably never will be able to obtain it in the future.

In the foregoing action, it was the attack alone which brought the certainty that the Russians had decided to retire no farther. Therefore, the attack had not been in vain. In a few hours it had clarified the situation to the high command. By no other means could this vital information have been so promptly obtained. Perhaps other methods might have succeeded, but only at the expense of several costly delays. Now, however, the Germans were in a position to concentrate all their strength at a single point to break through the front, to drive the Russian from his burrowings in the earth and once more bring about the return of open warfare.

Several days later, the battalion which we have been following found itself in the role of conducting such a decisive attack.

Again the Germans had come face to face with a Russian position and again they had launched an immediate attack against it. The Russians had selected a strong position which threaded in and out among innumerable small lakes, and which strongly resisted the movements of the attacker.

Our battalion had been assigned the mission of barring one of the narrow gaps between two of these lakes and engaging the Russians with fire. The attack was to be launched at another point in the line.

The hostile position lay on a hill. A field of fire extended about two hundred meters in front of it; immediately in front of this lay swamp and woods. Our battalion slowly worked its way forward through this marshy wood but no soldier showed himself on the further edge. By laborious work a few shallow trenches were built along the fringe of the wood and carefully concealed.

The Russian lines could be clearly seen on the hills to our front, their entanglements glistening in the sun.

During the afternoon there was heavy firing to our right and left and, in accordance with our mission, we opened fire without attacking. This fire was returned so energetically that we were certain that the hostile trenches were occupied in force. Meanwhile the enemy artillery searched the woods in which we were concealed without doing us any great damage.

Toward evening the noise of the battle died down and the thunder of cannon, which had prevailed all day, subsided. Patrols immediately went forward to determine whether or not the Russians had deserted their trenches, but they encountered fire on every hand and were driven back. It was patent that the enemy was still determined to defend his position. Again, the first German attack had failed.

All that the higher commander knew of the situation was the fact that his organization had run into a Russian position. He could not know whether it was a rear guard or whether it was the enemy in force. It might be merely another delaying action or it might be a determined stand to halt the German pursuers for once and for all. Clarity only by the attack—and the attack had to be launched quickly. If the opposition confronting the German advance were only a rear guard, then it must be quickly overcome in order to deny the enemy time in which to reorganize, for every day that he gained through a successful rear guard action was a tremendous advantage to him and a correspondingly great disadvantage to his pursuer. Similarly, if the advance had been halted by well organized lines of resistance, every day lost would permit this front to become stronger and stronger. It was essential to determine at the earliest possible moment the exact type of opposition that the Russians intended to offer.

This could not be accomplished merely by confronting the position and reconnoitering it with patrols. Against an entrenched enemy, patrols can determine little beyond the fact that this or that line is the forward limit of a defensive zone. They can never determine how strong the defensive position actually is. Neither can they gauge the tenacity with which the enemy will defend. No! The quickest and most certain way to cut the Gordian knot of war is to attack and, so far as our limited vision may reach into the future, it seems likely to remain the soundest course to follow when in doubt.

To digress for the moment, one of the most famous examples, in which an obscure situation was clarified by the attack, was the offensive movement of the IV Reserve Corps under General von Gronau during the battle of the Marne, in early September of 1914. The German First Army with its right flank to the east of Paris was advancing to the south. To protect this exposed flank the IV Reserve Corps was left behind along the Ourcq. The reconnaissance agencies of this Corps reported suspicious movements in its front. Wherever they turned, German patrols encountered an impenetrable enemy screen. As a result General von Gronau decided to cross the Ourcq with his entire corps and attack, in order to clear up the situation. He effected the crossing and immediately attacked. His corps, ripping through the hostile screen, uncovered the French Sixth Army on the point of launching a decisive attack into the German flank. In a single day the attack of the IV Reserve Corps had disclosed the entire scheme of the French. A little delay and the threat to the German flank might have materialized in disaster.

Returning to the campaign in Russia, the attack which had failed, confirmed the fact that the enemy was determined to maintain his position. The problem confronting the aggressor was to decide upon the best means of breaking through the Russian position. A penetration was agreed upon and the battalion with which we have been dealing was selected as one of those to make the main effort by piercing the hostile lines directly to the front. We are not here concerned with the factors that prompted this decision, but the proximity of a first class road which expedited the movement of reinforcements, and the forward displacement of artillery, undoubtedly played a major part in the high commander's decision.

Orders for the action arrived during the night. The attack was scheduled to jump off at four o'clock the following evening. Several other battalions which were to participate in the attack came up under cover of darkness.

At daylight the next morning several artillery officers appeared and erected their observation posts. Officers of the infantry unit they were to support immediately acquainted them with all known details of the Russian position. Above all, they were told where hostile machine guns had been located. Also they were thoroughly oriented as to the infantry's plan of attack, points on which artillery fire was particularly desired and positions along the hostile front which were considered especially dangerous.

Here is a point that will bear special emphasis. When all is said and done, it is the infantryman who makes the attack and in whose hands rests the decision of victory or defeat. It is his back which sustains the heaviest burden; his body which suffers the greatest hardship, and his life which is suspended by the most tenuous thread. Therefore, it is he who is most vitally concerned with the enemy, with the hostile position, with the terrain over which he must operate, with the artificial obstacles which he will have to overcome, and with a thousand and one schemes for the conduct of the attack. He strives to form a mental picture of the probable hostile resistance and of various measures to overcome it. His thought is something like this: "If our artillery will place a heavy fire on that hill, where a number of hostile machine guns are located, then the enemy will be unable to use those weapons. Further, if it will cut the enemy's wire at that point so we can break into his trenches, I believe our attack will be successful." So runs the thought of the foot-soldier.

The critical matter, then, is to instill in the attacking infantry the firm belief that it will succeed. If this can be achieved, victory is already half won. What would happen if the infantryman expressed his requests and intentions to the artilleryman, and received the reply, "No I am going to do it differently. You must organize your attack to suit me?" It would not be difficult to guess the result. His will and his viewpoint having been completely disregarded, the infantryman's belief in the success of the attack would suffer accordingly. He must always have the feeling that his viewpoint, with regard to the details of the attack, has been carefully considered, for every man believes strongly in the success of an undertaking if

[75]

it be executed in accordance with his idea. If, in these conversations, the artilleryman is capable of making the infantryman feel that it is the action and initiative of the infantry which is the paramount matter, then he has assisted that infantry almost as much as by the actual fire he will later deliver in the attack.

In this case the commander of the 9th Company and the artillery liaison officer, who came to discuss the plan of attack with him, knew each other from previous battles. Their trust was mutual. When the artilleryman informed him of the small number of batteries that had been allotted to support the attack the company commander was horrified. However, his friend reassured him: "You certainly know how well we fire and that up to now, no attack we have made together has ever failed. The Russian positions over there, that you wish us to concentrate our fire on, are so favorably located for my batteries that you may rest assured we shall smother them." The two officers then discussed the details of the approaching attack and both were convinced that it would be successful.

Having completed their plans with the liaison officer, the company commanders now discussed the attack with their platoon leaders, who in turn discussed it with their soldiers. There at the edge of the wood every detail was carefully gone over: "The 1st Platoon goes forward through this ravine and captures that little hill over there, then it attacks in the direction of that high tree which you can see standing up beyond the hill. The 2d Platoon advances on the left of the 1st Platoon to that hill and captures the two trenches on it which lay one behind the other. After that they capture those houses just beyond the hill. After the attack is launched the 3d Platoon advances to the line of departure and follows the 2d Platoon, as soon as the 2d Platoon has captured those trenches. The two machine guns that have been ordered to support us will fire from this corner of the forest until the assault platoons have entered the hostile trenches, then they will follow to that hill. I will remain, at first, with the 3d Platoon."

This done, the company commander then interviewed the commanding officers of the companies on his right and left in order to insure cooperation. A bit later the commander of the reserve company came forward with his platoon leaders to orient them on the terrain, and shortly thereafter the battalion commander appeared to discuss final arrangements with his company leaders. The two machine guns arrived and extra ammunition was distributed. Hot food was brought from the field kitchens. This was one of the most important preparations of all, for no one knew when the next meal could be served. Moreover, a soldier always fights better with a full belly.

These endless little details have been exactly portrayed in order that the reader may appreciate the myriad things that must be thought of and accomplished in preparing a coordinated attack. At the very beginning of the war, all armies learned that far more time is required to prepare an attack than had been thought. They also learned that once battle is joined the opportunity to issue detailed orders is gone. For this reason, orders should cover everything that can be foreseen. The mission, in particular, must be unmistakably clear so that once units become engaged all subordinate commanders can act with a unity of purpose.

It was now 3:40 PM and the German artillery preparation roared overhead. It lasted only twenty minutes but, to its great joy, the infantry saw the shells ripping into those points of the Russian position which they considered critical. The enemy replied in kind, energetically searching the swampy wood with artillery and machine guns.

Promptly at four o'clock the German infantry rose from their shelters and moved forward. Initially they encountered only a thin Russian fire. Everything appeared to be going well. Suddenly a machine gun cut loose from the right front. It appeared in a location that had never even been suspected. Moreover its position could not be seen from the artillery observation post. In a few moments twenty men had fallen, each killed by a bullet in the head. The advance of the 1st Platoon, or rather what remnants remained of the 1st Platoon, was immediately halted.

Seeing the attack of the 1st Platoon collapse, the company commander was now forced to decide whether or not he would commit his reserve platoon in order to further the advance of this portion of his line. Correctly, he decided against it. Had he committed his reserve at this time it would unquestionably have been decimated as it entered the zone of fire of this murderous Russian machine gun. Instead, he ran to his two machine guns located in the corner of the wood and directed them to cover the hills to the right front, which were the objective of the 1st Platoon. In this manner, he prevented the Russians in this area from acting against the 2d Platoon.

Meanwhile, covered by a slight rise in the ground, the 2d Platoon got forward quickly without sustaining casualties. In a few minutes it broke into the Russian position. A brief hand to hand fight ensued, a few of the Russians raised their hands and the Germans disappeared over the crest of the hill.

The time had now come: "The 3d Platoon follows the 2d Platoon through this ravine," yelled the company commander. Turning to the machine guns, he called out, "Follow as quickly as possible." He then ran forward. With him were four men, his true companions in many battles. These, his striker and three runners, comprised his entire staff.

Breathless, he reached the hill which had just been captured by the 2d Platoon. A scene of horror greeted him. Germans and Russians, dead and dying, were sprawled in the shell-torn trench. Where was the 2d Platoon? Its orders had been to attack straight to the front—but it was not to the front. Suddenly he saw it fighting off to the left front. It had allowed itself to be diverted by the action of the 5th Company on the left, and since the attack of the 1st Platoon had been stopped, a nasty gap had developed in the German front.

At this moment the company commander's attention was abruptly diverted to the right rear by the sound of heavy fire. He looked back. The Russians, on the hills which the 1st Platoon had been attacking, being no longer bound to their trenches by the two German machine guns were launching a counterattack. He saw the 3d Platoon become involved in this fight.

At this point one of the runners called out: "The Russians are attacking our hill!" Noting the gap in the German line, the enemy had launched another counterattack, this time to the front and against the hill on which the company commander and his four soldiers were standing. Coincident with this, heavy machine-gun fire began to sweep the hill. Two runners were immediately hit; one was instantly killed and the other, mortally wounded, died the same evening. The company commander and his striker opened fire on the storming Russians while the last runner went back to the 3d Platoon for reinforcements.

The situation was critical. This hill was a key position. If the Russians succeeded in taking it, not only would the local German attack be shattered but the troops attacking on the left would probably be destroyed, for this hill commanded the battlefield as far as the lake. Therefore, there remained nothing for the company commander to do but stay on the hill, fight, and hope against hope that portions of the 3d Platoon would arrive in time.

The Russians, who always attacked with courage, closed up to eighty meters, then seventy meters, then fifty meters. The two Germans fixed their bayonets with the intention of selling their lives dearly. Suddenly one of the two machine guns of the 9th Company (the other had been destroyed by artillery) began firing. It had gotten forward at last. At pointblank range it let loose a hail of bullets among the Russians, mowing them down as with an invisible scythe. The counterattack immediately collapsed. A few Russians who tried to fly were promptly shot down. A few others, drawing on their last reserve of courage, attempted to attack, and these, too, were killed. Eighty survivors lost courage, threw away their weapons and surrendered. These became German prisoners.

Despite this success, the crisis had not yet passed. The 3rd Platoon was still fighting far to the rear. From the hill on which the company commander was standing, the progress of the fight could be watched. The machine gun now turned around and opened up to the right rear, taking the Russians in flank. In a few moments the outcome of the fight on this portion of the field was also decided. The 3rd Platoon was freed and came forward to join the company commander. Meanwhile the hill for which the 1st Platoon had been fighting was evacuated by the enemy.

The company commander now continued the attack with

[79]

the 3rd Platoon and the remnants of the 1st Platoon. They stormed and took the weakly defended houses on the next hill. Just behind the hill they saw four Russian field pieces on the point of being withdrawn. Again the faithful machine gun opened fire. The horses fell and the battery became German booty. The Russians unwilling to relinquish their artillery so easily, counterattacked and recaptured the battery, but at this point the 5th Company and the 2nd Platoon of the 9th intervened in the fight and for the second time the Russians were forced to abandon their guns.

Gradually it became dark, but the Germans continued their attack a number of kilometers, exploiting the success of their breakthrough. The Russian position had cracked Open warfare was restored.

The German losses had been heavy. The 9th Company, for example, which had entered the action with one officer and ninety men, emerged with one officer and forty men. In this one engagement they had suffered more than fifty per cent casualties.

During the night the wounded were taken back and the field kitchens came forward. It was a bit of luck that mail from the homeland arrived that same evening, but there were many letters and greetings for the 9th Company that could not be delivered.

And so we reach an end. In this article we have pictured several phases of combat in that borderland that exists between open and trench warfare. We have seen the difficulty of determining whether the enemy is bent on decisive battle or is only seeking to delay. We have emphasized the fact that the attack alone can gain that immediate clarity so essential to the aggressor if he is to avert stabilization. We have pointed out that troops must know and constantly keep before them the truth, that such attacks are absolutely necessary even if they fail. And finally we have seen the tremendous burden the infantry must bear in such battles.

From this action certain lessons that have already been discussed in earlier articles may be drawn. Their importance warrants a restatement:

1. Attacks must be well prepared and discussed if they are to succeed.

2. There is only one opportunity to issue detailed orders

and that is before battle. When the action has actually begun, orders must be short and simple.

3. Every fight develops differently than is expected. Officers and troops must realize this in peace, in order that they will not lose courage when the unexpected occurs in war.

4. Little is known of the enemy in war. The attack is the best way to dispel this obscurity.

5. Reserves must be employed at that point where troops have been successful and not where they have failed.

6. And finally, it is again seen that success and a knowledge of battlefield psychology are powerful influences for inculcating a sense of confidence and security in troops. Nothing will strengthen their ability to resist the disconcerting impressions of battle more powerfully than these.

CHAPTER VIII

The Development of the German Army Since the World War

The events about to be related constitute one of the darkest periods of Germany's entire history. It is not pleasant to tell how a nation, after four years of war, broke apart and its people fell to fighting among themselves. America knows that civil war is bloody and terrible. Civil war came to Germany after that nation had lost a war in which it suffered untold bloodshed and while it lay destitute and starving. And yet, the manner in which soldiers helped overcome these terrible conditions should interest military students when the facts are laid before them by a brother soldier.

Revolution broke out in Germany on November 9, 1918. The Armistice was signed on the 11th. The German army streamed back to the fatherland and in a short time was demobilized. In a few weeks there was no army. All discipline had vanished. As in every revolution, the idea of liberty was predominant; everyone thought he could do as he pleased. The result was that no one wanted to serve any longer. Everybody wanted to command or to acquire a lot of money without working for it. In industry, workers demanded higher wages and shorter hours and strikes became as common as the need of daily bread.

At this time several factions in German were battling for control. The revolution placed the Socialists in power, but the Bolshevists wanted to rule Germany and transform it into a Russian Soviet. Both of these factions battled not only with words but with weapons. In one part of Germany the Socialists were predominant, in another the Communists. In every German state and province, in every town and village, there was dissension and civil strife. Bandits took advantage of these conditions to rob, plunder, and murder.

With Germany in such a condition, millions of young men without other employment, but well trained in the use of arms, offered their services to the various factions to fight for control of the government. Weapons were available everywhere. Officers were chased home and not allowed to come out. Whenever they were seen on the streets, they were humiliated, insulted, even attacked and beaten. Many were killed. In addition to all this, the Poles in the east, without waiting for the outcome of the Armistice, invaded German territory in Silesia, while Bolshevists threatened our eastern boundary. Such were the conditions in Germany on Christmas, 1918, when all other people were celebrating the great day of "Peace on earth, good will toward men".

The government, now unable to defend itself against the activities of the Communists, called for the formation of a volunteer army. Volunteers had already formed under Hindenburg, who had offered his services to the fatherland, and these had prevented further invasion by the Poles. This force was supplemented by those officers and non-commissioned officers in nearly every small garrison who had remained loyal to the government.

On the other hand, many garrisons that were well stocked with arms, ammunition and clothing, were occupied by bands of young men under command of Communist leaders who declared that their new military power was the government. Although they performed no duty they received pay. Since they had arms, they terrorized the people. For instance, it frequently happened that they would go to a bank, demand and obtain a large sum of money, and then divide it among themselves. The best talker became their leader. They took no orders from the government but did as they pleased. They ruled with the fist. For them the new freedom meant pay without work.

As a result, the people hungered and starved. For several years there had been a great scarcity of milk, coffee, eggs, and tobacco. I remember that even as late as the year 1920 we often went to bed early because we were hungry. Each person received per week but one loaf of bread, one-quarter pound of butter and one-quarter pound of meat. We ate the butter and the meat on Sunday, together with a slice of bread. Even now when I think of these times I hardly know what we really lived on.

In February, 1919, volunteers were called. Nearly every

officer proffered his services. New commands were formed. There were plenty of volunteers, because therewere still plenty of people who wished to overthrow Bolshevism, and also because many people had not been able to find work. Unfortunately, there was no general policy to guide this movement. The government needed the officers to get out of its difficulty but, on the other hand, there was fear that the officers might turn about and destroy the government. Therefore, it established political military commissions to watch the officers. These tried to interfere with officers who were carrying out their orders. Frequently the military commissioners endeavored to oust the officers and secure control themselves. Gradually, however, the officers managed to secure the command of their own units.

Let it not be thought, however, that we now had excellent troops. The soldiers served for three months, but they had the privilege of leaving their units after giving 14 days' notice. This was often done when they learned that they were about to engage in a fight. Therefore, there was a continual going and coming in units. Sometimes companies of 150 men would be reduced in a few days to 30 or 40 men. Sometimes the men decided that they didn't want this or that officer and therefore refused to serve. If a commander were to be changed, the soldiers wanted to be consulted about it beforehand.

Many towns and provinces refused to have such government troops. They established their own troops according to their political sympathies; if they were Bolshevists they had troops of Bolshevistic tendencies.

These new units differed entirely in strength. In one place a company was formed, in another a regiment, etc. Sometimes a regiment consisted of two battalions, sometimes of three battalions—sometimes with machine guns or trench mortars, sometimes without. The soldiers still wore the old uniforms, but each unit had a distinct insignia. The prettier the insignia, the greater the number of soldiers who joined. Sometimes a commander with financial backing was able to pay his soldiers more than other leaders. Naturally, many men came to him.

These new units began to preserve law and order in their own immediate neighborhood. Sometimes they were able to do this without any trouble, but more often than not there was severe fighting. The hardest fighting took place in Berlin. Artillery and trench mortars were employed for days. Some of the Berlin companies consisted entirely of officer personnel. Later we grouped these troops into larger units and formed reinforced brigades. These brigades were then employed against the states and towns that were in revolt. Long campaigns took place against Bavaria, Saxony, Thuringia, and Braunschweig. Meanwhile, other units fought on the eastern boundary against the Poles in Lithuania and against the Russian Soviets in Courland.

Only with the greatest difficulty could the restless masses be held in check. Time and again uprisings occurred and time and again troops had to be employed. Moreover, difficulties arose within the units themselves. Frequently the troops attempted to force the officers out of their commands. Conditions were terrible in Germany as the year 1919 came to an end.

In the meantime, the treaty of Versailles was signed. This stipulated that our army should be limited to 100,000 men, including 4,000 officers and our navy to 15,000 men, including 1,500 officers.

In the abrupt creation of this new army, thousands of officers and practically all of the enlisted personnel had to be discharged. New soldiers had to be found who were willing to pledge themselves to a long enlistment. Meanwhile, there still remained the task of quelling new uprisings.

In March, 1920, counterrevolution broke out. An attempt was made, through a new revolution, to destroy the republican form of government and reestablish a monarchy. Under the monarchy the army had been strong and powerful and the officers had held a respected position. One of the first acts of the republic had been to disband the army and most of the officers had been chased home. It is not surprising then that a large number of officers and their entire commands joined this new revolution. Some of them, however, saw that such an attempt could not succeed, because the people, as a whole, did not want a monarchy. Therefore they remained loyal to the existing government. Others—and I believe they were by far the greatest number —did not bother themselves about politics at all. They saw that it was their duty to restore peace and order so that a peaceful development might be possible. These also stood by the government.

The Bolshevists utilized the opportunity afforded by the counterrevolution to establish their power and it was not long until they controlled the great industrial centers. They organized Red troops. There was no dearth of men who understood the use of weapons. They established rule by terror after the pattern of the Russians. Russian money and Russian leaders were sent to assist them. During these trying days, it seemed that Germany would surely disintegrate. What a situation! We had a republic, then came the counterrevolution that attempted to reestablish the monarchy, and against both of these stood the Red Communists. The one stable factor in all this chaos was the army. For the second time the country was saved by her soldiers but above all, by her officers. The counterrevolution failed in a few days, but the Reds made full use of this brief period to recruit their own ranks. A long campaign now began against them.

The largest and most important industrial center of Germany is the Ruhr. Without it, Germany cannot live. This entire section was in the hands of the Reds who were attempting to overthrow the state, using this area as a base. There were but few available troops that could be sent against them. Troops from all parts of Germany had to be assembled, but this concentration could not be effected until peace and order had been established in their own localities.

Divisions were finally assembled and the Ruhr Valley was besieged on all sides. For a time the situation was critical. There was severe fighting and the losses on both sides were heavy. In a fight in one of the large towns one of our regiments lost ten officers dead and wounded. The fighting continued until May. After May peace was gradually restored. Only an occasional uprising occurred here and there that the troops would have to put down.

Throughout this period the discharge of troops continued. During the year 1920 the enlisted trength varied between 200,000 and 150,000. Many of the volunteer corps that had helped the government had to be disbanded. Some refused to disband and had to be forced to. Sometimes soldiers had to be sent against their own comrades. Some of the volunteer corps commanders had been either captains or first lieutenants. In the new regular army, these places had to be filled by older officers. Frequently the soldiers refused to recognize a new commander. Some of the volunteer corps commanders would not accept demotion. Out of such situations an unending number of difficulties arose all of which had to be surmounted. However, the required demobilization was finally accomplished.

To add to the complications of reorganization, no one knew anything about his own future. Regimental and brigade boards were established to determine which officers should be retained in the new army. These boards included young officers, as well as commanders. They decided which officers should be discharged and which should be retained. To-day we know that they performed their work well.

Thus ended the year 1920. On the first of January, 1921, the new army of 100,000 men was to be put into effect. For a short time after January there was about 50,000 additional troops but these were all discharged by the end of April, 1921. With this the new army was complete but complete only to the same extent that a human being is complete when first born. It still had to learn to walk, talk, and think. This required time, peaceful development, and ample nourishment, but all of these elements were lacking.

The nourishment of an army consists of weapons and equipment. In this new army, we have no artillery, no air service, no tanks ,and no chemical warfare service. We are permitted to have only two uniforms for every soldier. The number of machine guns and minenwerfers—yes, even the number of cartridges—are all definitely prescribed. For example, the entire bridge equipment of our seven pioneer battalions is sufficient to build only one bridge 500 yards long. Such a lack of equipment makes even the thought of engaging in a modern war untenable. Moreover, training with such limited equipment requires an unusual amount of imagination; theory has to substitute for practice.

To make matters still more difficult, we have had little opportunity for peaceful development. In the year 1921 several regiments had to be sent to the center of Germany to put down a revolt there. In fact, until the year 1924 many uprisings occurred that had to be quelled by troops. Then the inflation occurred and all previous ideas concerning the worth of money were destroyed. All money seemed worthless; the money of one day would buy nothing the next.

Then came the winter of 1923-24. The French covered the Rhine and occupied the Ruhr Valley. In Saxony and Thuringia the Communists again ruled. In Bavaria a new attempt was made to stir up a revolution. The Poles invaded Silesia. Inflation reached its climax. Just as the army was beginning to develop morale, it was again called upon to suppress disorders. Fighting continued throughout the entire winter until the following March. Then order was again restored. Time and again, when the nation was in trouble and danger, the soldier was called upon and each time he saved the day.

Since 1924 the army has been able to develop under relatively peaceful conditions, but not without great difficulties. Even to-day many of these difficulties have not been overcome. Consider the limitations of our army and the problems that these limitations entail. For instance, the soldier has to serve twelve years. It is exceedingly difficult to offer the soldier progressive instruction for twelve years and maintain his professional interest. He must also be fitted for an occupation after he is discharged. He comes into the army when he is eighteen or twenty years old. This means that he will be thirty or thirty two when discharged. He is too old then to start learning a trade or profession and too young to remain idle the rest of his life. Therefore, schools have been established which the soldier may attend during his enlistment in order that he may prepare himself for an occupation of his own choosing after discharge. One disadvantage of this is that during his enlistment he may think too much about his future occupation and lose interest in his duty as a soldier.

In such an army with its long enlistment period, it is natural that much is demanded of the officers. Discipline, which was lost during the wild years of the revolution, had to be slowly reestablished in the face of difficult internal and external conditions. New regulations embodying the lessons we learned during the World War had to be prepared and distributed. Not only did the troops have to be organized, housed, and equipped, but at the same time they had to be taught entirely new tactics. The rich war experience of the officers and non-commissioned officers was a great asset but their task was a difficult one. In looking back upon the first ten years of our new army, we see reason to be proud of our accomplishment. Everywhere our army is spoken of with considerable respect. It is small, but it is efficient. Discipline and morale have been restored. The professional ability and standards of both officers and men are high. All this has been accomplished under considerable difficulty and within a short period of time. We have built ourselves a new house and this house stands firm and safe.

CHAPTER IX

The Army of the United States

(*This article appeared in the January 25, 1932 number of the Militar Wochen Blatt.*)

Every army must be an organic element of both its people and its state if it is to fulfill its mission. If we hope to grasp the spirit of an army, if we seek to understand its opinions, its thoughts, its organization, we must first concern ourselves with the mission that is assigned by the state.

The active (regular) army of the United States is about 120,000 men strong and has about 12,000 officers. That is certainly not a great number for this huge country, particularly when we consider that a large portion of this army is garrisoned in Panama, Porto Rico, Hawaii, the Phillippines and China. If we examine an atlas we will find that Germany and the United States appear on identical pages. Therefore it is difficult to arrive at a comparison of relative sizes from this source. For that, we must turn to a globe. Picture this—the United States is composed of forty-eight separate states, the largest of which, Texas, has an area one third greater than all Germany. Its population, however, numbers only about five million inhabitants. If this single state were as thickly populated as Germany, it would have from eighty to ninety million inhabitants. But in all America, there are at the present time not more than one hundred and twenty million people.

An express train runs from Cologne to Konigsberg, the entire width of Germany, in twenty-four hours. The fastest express requires three and a half days to cross the United States from New York to San Francisco. In Europe, such a journey almost corresponds to a trip from Lisbon to Moscow.

In this immense land there are roughly one hundred thousand soldiers. Naturally they are widely scattered. One hundred thousand soldiers in such an extensive area seems a small number indeed according to our standards. Does such a small army suffice for America?

In comparing military strengths the average American citizen naturally generalizes on the basis of his country's situation. He asks the foreign visitor—"Why do the European states maintain such large armies? Why, for example, is Germany not content with one hundred thousand men; she is much smaller than the United States?" Quite apart from the fact that the United States has reserves at its disposal who are more or less trained, while Germany, as a result of Versailles' dictation, has none, the answer is that conditions are different, strikingly different. The United States is not a country in a European sense, but rather a continent. On account of their world political situation, the country is not dangerously exposed to an attack by land.

The American doesn't need to worry about a sudden invasion. He doesn't have to let his enemy dictate the moment when he must defend himself. He reasons: "To-day I'm not ready to go to war, but I shall be later on. I shall not be ready to fight for six months, therefore I will not start fighting until then. I, not the enemy, will decide the day I shall throw my army into battle. I shall always have just as much time for preparation as I consider necessary" How different the situation in Europe!

America does not need a regular army in proportion to its size to insure its national safety. Other motives are decisive. In the event of mobilization the small regular army will constitute the nucleus about which the enormous emergency army will form. The regular establishment is a cadre army whose principal mission in peace is the training of reserves. It is not so much an army to be brought to full strength in case of war, as an instructional personnel for that war army which will first come into existence upon mobilization. We therefore find its chief mission in peace summarized in the single word—instruction.

The fundamental difference between the American army and most of those of Europe is that in America, the role of teacher occupies the foreground of attention. That is certainly a tremendous task to be accomplished by a mere twelve thousand officers in an army of millions scattered over a wide expanse of territory. In time of peace, these millions of potential soldiers are civilians who follow their professions and trades. They do not serve for a definite time in the ranks of the active army as was the case in pre-war Germany and as is still the case in most European countries to-day. To be sure, they train, study, prepare, and educate themselves during peace, but only in so far as their civil callings permit. In other words, the military is secondary. When war breaks out, these troops are not commanded by the regular officers. They take the field under the command of the civilian leaders who have had direct charge of their training in peace.

Under such circumstances, what does the American regular officer do? His chief role is that of teacher and counselor of the great war army. For the larger part of his military career he serves in instructonal capacities far from the regular troops and entirely on his own.

The training of the regular officer must naturally conform to his principal activity. In order that the officer may be released as quickly as possible for instructional work, his training is given in condensed, compact courses. How could such training be given to officers in regiments? In our army, we know no other way of training officers than with troops. All other means—schools or courses—are purely supplementary. In America, however, this constitutes the fundamental feature of an officer's training.

The American regular army is scattered over that enormous country in countless little garrisons that are often separated by several days' travel. If such conditions prevailed in Germany, it would mean that a regimental commander might have to travel from Danzig to Konstanz in order to visit all his companies. Such cases are not infrequent in America. Generally the garrisons lie outside the cities on federal ground and are known as "Forts" or "Camps". The development of the nation and the use of the army in the preceding century as outposts against the Indians compelled this splitting up of units. In these detached forts, the troops are dependent on themselves for everything. This means that they build and maintain roads, paths, and, in part, the buildings. With the personnel of these small garrisons thus withdrawn to a large degree from strictly military duty, an orderly and useful training is difficult. In addition the

American Constitution restricts the use of other than federal ground for training and prohibits the billeting of soldiers in civilian houses. The result is that comprehensive terrain exercises or maneuvers can be carried out only at those garrisons which possess extensive training areas and such garrisons are few. The occasional transfer of troops to suitable training areas is at best a poor substitute for this deficiency in their own garrisons. Moreover, these transfers are difficult in view of the great distances involved. Still another factor which militates against the most efficient training of combat organizations is a shortage of officer personnel.

It is obvious then why officers can not be trained to become first-class teachers during the comparatvely short period they serve with troops. In this period, too many different duties absorb their time. It naturally follows that this training must be given in schools. The curricula of the schools are based on the idea of equipping the officer with everything he will need in his instructional capacity. He learns the more important concepts, both from a tactical and a technical viewpoint. He is also taught pedagogy and the best methods of instruction in military subjects. Service with troops is no more than an interlude in this constant change between teaching and being taught.

Thus we see that the American regular army system provides for enormous educational institutions devoted principally to developing teachers who will train the reserves and the nation at large. It is truly a fine system—but we, bound in our chains, can only look at it with envy.

It does, however, have its disadvantages. To be an officer means to be a leader—to be a leader of troops in battle. It is certainly correct that leaders, like great artists, are born and not made; but even the born artist requires years of hard study and practice before he masters his art. So it is with the military leader; if he is to learn the art of war, he must practice with the tools of that art.

As previously stated, the American officer's primary function is that of a teacher since he must teach others how to teach. But to be a successful teacher must he not be a leader as well? It should be remembered that the regular army will also participate in war and under its own officers. But if the American officer is primarily a teacher with his principal training in schools, does sufficient time and opportunity remain to develop him as a leader?

Within the limits of this brief article we can only suggest answers to such a question and the undoubted difficulties attendant thereto. How is a uniform viewpoint of the officer corps attained? Under the conditions prevailing in the Army of the United States it is obvious that the officers' schools cannot be content with introducing their students to the world of military thought and merely guiding their thought processes so that their intelligences will ripen later under the guidance of older officers. Since the opportunities for practical training in the service are limited the student must take away with him from the schools definite answers to his questions; he must be given standard ideas and principles. His thinking must be led down prepared channels to the solution recognized as most appropriate. Independent thinking must take second place to a uniform solution. Therefore, the danger naturally arises of too much emphasis being placed on set forms.

Now for an answer to the other question I raised: the double training as teachers and leaders. We have sought to make clear that the principal task of the American officer is teaching—far distant from the troops—and that the knowledge essential for this purpose is given in condensed courses in schools. The Americans realize, of course, that the officer must be trained as a leader as well as a teacher. They also recognize, with regret, that the achievement of the ideal, i. e., of training teachers and leaders at the same time, is unattainable owing to human imperfections and the short period of the school courses. Nevertheless, everything possible is done to train the students as leaders. War strength units are at the disposal of at least one of the schools (The Infantry School) and all leaders' posts in these units are frequently filled with students during combat exercises.

Once in each grade every officer, as a rule, performs command duty for two years in the foreign possessions. The units in these possessions are stronger numerically than corresponding units in the United States and offer the best opportunity for training officers as leaders. Nevertheless, the emphasis of both the activity and the training of the American regular officer—it is repeated once again—is in the teaching field.

Before the war the German army, in addition to its other duties, was charged with the military training of the people as a whole. We solved our problem by training reserves within the regular army. We could do this, because the numerical strength of our army permitted it. The numerically weak American army cannot train these reserves in its own units, even if historical and political grounds did not forbid it. Other ways had to be found. Their system seems strange to us but it is undoubtedly best suited to American conditions. The relative geographic isolation of the United States renders their scheme altogether practical. These American methods, however, are neither suitable nor practical in Europe without material modification.

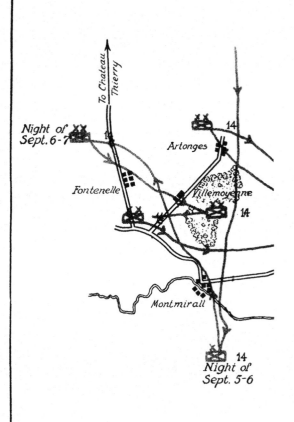

To Chateau Thierry

Night of Sept. 6-7

Artonges

Fontenelle

Villemoyenne

Montmirall

14

14

14

Night of Sept. 5-6

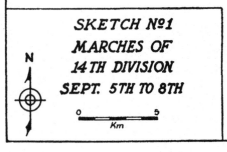

SKETCH Nº 1
MARCHES OF
14 TH DIVISION
SEPT. 5TH TO 8TH

N

0   5
Km

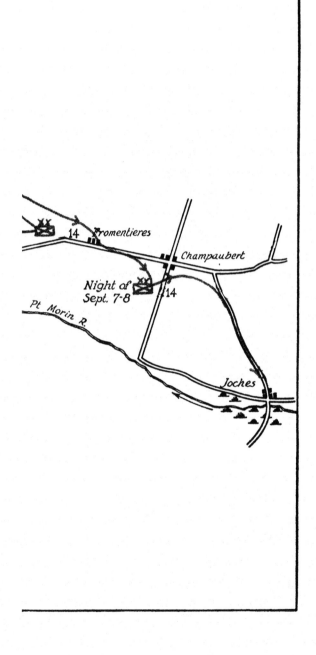

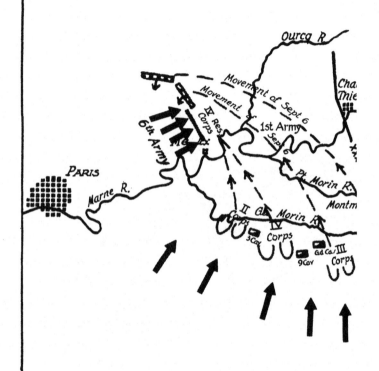

Ourcq R.

Cha
Thie

Movement of Sept. 6

Movement

IV Res
Corps

6th Army

1st Army

Sep. 6

Me. Li.

PARIS

Marne R.

Pt. Morin R.

Montm

II Gd. Morin R.
Corps

IV
Corps

3 Cav

Gd Ca./III
Corps

9 Cav

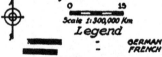

Situation
1st -2nd-3rd German Armies
Sept. 5 1914

0   15
Scale 1:300,000 Km

Legend

GERMAN
FRENCH

SKETCH № 2

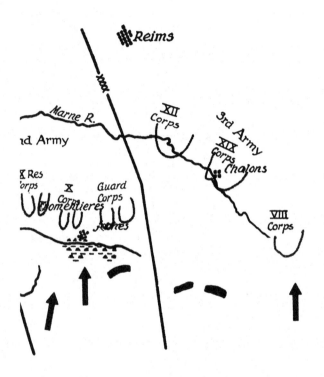

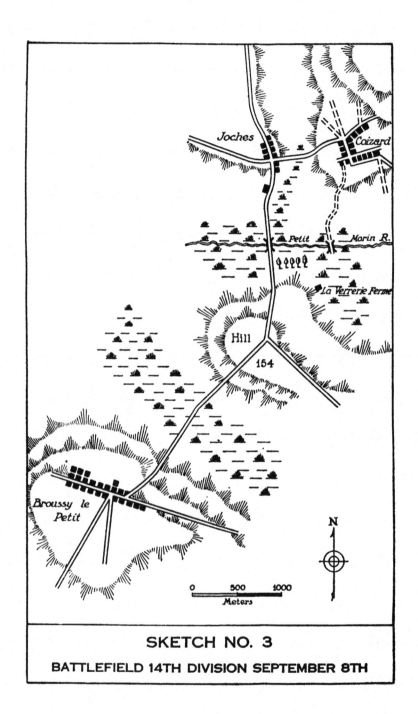

Joches

Coizard

Petit   Morin R.

La Verrerie Ferme

Hill

154

Broussy le
Petit

N

0   500   1000
Meters

SKETCH NO. 3

BATTLEFIELD 14TH DIVISION SEPTEMBER 8TH